Wisdom of the Body Moving

An Introduction to Body-Mind Centering

Linda Hartley

North Atlantic Books
Berkeley, California

Published by
North Atlantic Books
Berkeley, California

Cover and book design by Catherine Campaigne
Illustrations by David Philbedge
Photography by Udo Hesse
Nature photographs by Simon Ferguson
Additional photographs by Katya Bloom and Linda Hartley
Printed in the United States of America

Body-Mind Centering™ denotes a patented system of movement therapy created by Bonnie Bainbridge Cohen. For purposes of the clarity and design of this book, the trademark symbol has been omitted from the term "Body-Mind Centering" when it appears in the text. However, this term is a registered trademark owned by Bonnie Bainbridge Cohen and is fully protected under U.S. law.

Wisdom of the Body Moving is sponsored and published by the Society for the Study of Native Arts and Sciences (dba North Atlantic Books), an educational nonprofit based in Berkeley, California, that collaborates with partners to develop cross-cultural perspectives, nurture holistic views of art, science, the humanities, and healing, and seed personal and global transformation by publishing work on the relationship of body, spirit, and nature.

North Atlantic Books' publications are available through most bookstores. For further information, visit our website at www.northatlanticbooks.com or call 800-733-3000.

ISBN-13: 978-1-55643-174-6

Library of Congress Cataloging-in-Publication Data

Hartley, Linda, 1953–
    Wisdom of the body moving : an introduction to body-mind centering / Linda Hartley.
        p.   cm.
    Includes bibliographical references (p.   ).
    ISBN 1-55643-174-0
    1. Movement therapy.   1. Mind and body.   I. Title.
    RC489.M66H37   1994
    615.8'2—dc20   93–44829
                                                                          CIP

11 12 13 14 15 EBM 20 19 18 17 16

I dedicate this book to the memory of my father—
the writing of it was his gift to me;
and to my mother who gave me the gift of life.

Contents

Movement Repatterning

The Body Systems

Conclusion

Acknowledgments

I would like to express my deep appreciation to Bonnie Bainbridge Cohen, the originator of Body-Mind Centering, for the unique work that she has developed, for the years of study and research that she has tirelessly pursued, and for the openness of heart, the generosity of spirit, and the inspiration with which she continually shares her work and vision. From her I learned to trust myself, my own experience and perceptions, my limitations and my gifts. I am indebted to her for the material which forms the basis of this book. I would like to thank her also for her invaluable comments on the manuscript and for her support of this project. My appreciation also goes to Leonard Cohen, codirector of the School for Body-Mind Centering, without whose hard work and dedication the school could not have flourished as it is doing today.

My thanks also to all the teachers, colleagues and students of Body-Mind Centering who have shared this journey of discovery with me and who have helped and supported my own learning in so many ways; and to all of my students and clients, from whom I have learned so much over the years, for the love, the courage and the enthusiasm which they have shared with me.

Without the support and encouragement of my friend and colleague Katya Bloom I don't think this book would ever have reached completion. My very special thanks to Katya for being there through it all with her love, friendship and enthusiasm for the project.

Thank you also to the many other friends and colleagues

who have supported and helped me in various ways throughout the writing of this book.

My appreciation for the wonderful work they have done to David Philbedge for the illustrations, and to Udo Hesse, Simon Ferguson, and Katya Bloom for the photographic work.

For appearing in the photographs I am grateful to Renate Deiss, Daniela Herlyn, Ute Lang, Regina Rudiger, Sygun Schenck, Barbara Schmidt-Rohr, Veronika Wiethaler, and Joachim Witzke. Thank you also to Carl Schmidt-Rohr, Megan Warner, and their parents.

The manuscript was read at various stages of completion by Bonnie Bainbridge Cohen, Katya Bloom, Dr. Jenny Goodman, and Sharna Travers-Smith; I would like to thank them for their invaluable comments and encouragement. Thank you also to Marianne Dresser and Lynne Uretsky for their meticulous editorial work.

And finally, my deepest appreciation to my spiritual friend and heart teacher, Lama Chime Rinpoche, whose continual love and wise guidance has sustained me throughout.

Foreword

Over thirty-five years ago I began an intuitive exploration of the relationship between the body and the mind. Through the years I have organized and systematized my insights and discoveries. The result has been the creation of Body-Mind Centering™ (BMC), a fundamental approach to embodiment through movement, touch, voice, and mind.

Thousands of people have joined in this process of exploration, but only two hundred or so have committed themselves to its full study and transmission. Linda Hartley has been a key person in the teaching and dissemination of this work in Europe. With this book, Linda has met the challenge of communicating her experience and interpretation of BMC with clear integrity and sensitive intelligence. *Wisdom of the Body Moving* is beautifully written, poetic and concrete, and in tune with the essential nature of the intuitive process underlying this path of embodiment.

Thank you, Linda.

Bonnie Bainbridge Cohen
Founder and Educational Director
The School for Body-Mind Centering™
February, 1995

Author's Preface

Body-Mind Centering is essentially about following the courses of nature, and so the work itself is a natural process of continual evolution. In a sense, this book presents a moment in a continuum of research and learning, a momentary point of crystallization of certain ideas concerning the development of human consciousness. Like all natural processes, the forms of expression that this body of work have developed continue to crystallize, dissolve, and evolve further as new insights, connections, and structures emerge.

The original draft of this book was completed in February 1989, and since then it has undergone some minor revisions to incorporate some of the more recent developments in the work. My own work has also developed during the years since I first conceived of and began to write the book. I now integrate into my practice and teaching of Body-Mind Centering perspectives from various areas of study, research, and personal experience with which I have been deeply involved over the years. Yet the essential principles and practice of Body-Mind Centering, described here, remain an important, profoundly effective, and valuable foundation for my work.

Body-Mind Centering has wide applications. It offers no fixed rules and procedures but demands that the practitioner or teacher draw upon her own creativity and personal experience in a way that will be unique for each individual. The practitioner is asked to respond from her direct experience of the moment, and of the particular circumstances of the individual or group with whom she is working. To support, guide, and orient the unique and evolving process of the moment, Body-Mind Centering offers principles and practical techniques, as well as language and theory based on authentic body experience. It gives a ground based in natural and organic processes from which we can each grow—both inward toward a deeper experience of our inner self, and outward toward expression of our unique self in the world. The work, as ground, also changes and evolves as we do.

I sincerely hope that you, the reader, will find in this book validation, support, and inspiration for your own growth and learning.

Linda Hartley
Cambridge, England
October, 1994

Introduction

The heart of this book contains a search for the wisdom that we all possess within us, the awareness of who we are on this earth. By journeying both deep inside to our own experience and out through our perceptions to the world we live in, we can begin to see who we may truly be, beyond conditioned self-images and habitual patterns of thinking, moving, and living.

The medium for our research will be the body and its movement. All that lives has the ability to move based on some personal motivation, whether conscious or unconscious, organic, instinctual, or volitional. Even a plant, as it grows, adapts its shape and position so as to be touched by the light of the sun. Movement in all its variety of forms is an expression of life and is essential to the continuation of life. As I live, I express my life-force in movement; as I move I feel my aliveness. To continue to live I continue to move and change. This life-force moves through us and expresses itself in the breathing of the smallest cell, the unconscious and conscious, subtle and gross movements of the body, as well as in the sounds we voice or the thoughts we think. Where movement is both free and integrated, there life will be felt to flow freely and strongly. This freedom and integration is a gift of our journey; as we explore in detail this complex and wondrous body, our life-force and thus our essential wisdom become more available to us.

Body-Mind Centering, the work that I describe in this book, offers a perspective and some techniques and guiding principles to help us in our research. The principles of Body-Mind Centering are based on the natural development and unfolding of potential within the human being. They describe the process by which we learn, transform, and recreate ourselves into new awareness and new form, from moment to moment. In this, we follow nature's cycle of birth, death, and rebirth. Body-Mind Centering concerns itself with the potential for growth, learning, and change inherent in each moment of experience. The *I Ching,* or "Book of Changes," speaks to the process of change and rebirth in describing the "Turning Point," the time of change at midwinter:

> The time of darkness is past.... After a time of decay comes the turning point. The powerful light that has been banished returns. There is movement, but it is not brought about by force ... the movement is natural, arising spontaneously. For this reason the transformation of the old becomes easy. The old is discarded and the new is introduced. The idea of RETURN is based on the course of nature. The movement is cyclic, and the course completes itself.... In this way the state of rest gives place to movement ... everything must be treated tenderly and with care at the beginning, so that the return may lead to a flowering.[1]

This cyclical rhythm of rest and renewal exists everywhere within us, from the life of the cells to the flow of our breath to our daily cycles of activity and sleep, and connects us deeply with the earth and all life forms. Through the work presented in this book, we can inquire within the body, acknowledging and strengthening our own internal rhythms and their connection to universal patterns.

This is a human story, a story not simply to read or hear, but to enter into, body and soul, with all our feelings and imagination. In this way we can experience the tones and shades and subtle shiftings that take place in ourselves in this unique and individual journey. This is learning of a different order, and in the process I describe here, there lies an opportunity to explore and connect with both the wisdom of the body and the knowing of the mind in a very direct and personal way.

This story is about embodiment, the human being at home, each of us in our own body. To be present in our body is a form of awareness, and it is a first step toward being kind to ourselves and others. In coming into our body we become connected to our greater home, the earth; we become a part of the earth and she a part of us. We are received into her, and she into us; we grow through and from her support and nourishment, and we express her qualities through our very being. She is our ground.

Inseparable from this "earth" aspect in the human story is the mind, or consciousness. In Chinese philosophy, consciousness is the "heaven" aspect. In the terms of ancient Taoist philosophy, earth and heaven, *yin* and *yang,* body and mind, coexist harmoniously in us. This is the way of nature, the Tao, the way of being fully human.

Too often, perhaps especially in modern Western culture, the union of body and mind, of the "earth" and "heaven" principles, is not harmonious. One is often overemphasized at the expense of the other, or one aspect may be denied, causing the other to suffer from exhaustion and distortion. The mutually enhancing connection between the two is lost and a sense of dislocation and disease ensues. We do not feel comfortable, "at home," in our body. Instead of knowing where we are, we feel lost and rootless. This is a fundamental source of the sickness of body and soul that many of us experience. As acupuncturist Dianne Connelly says, "All sickness is homesickness."[2] And so we search, attempting to return to our knowledge of who we are and where we are.

It is part of the human condition to search, not knowing exactly what it is we seek, yet somehow sensing something hidden in our hearts. This mysterious balance of "knowing" and "not-knowing" drives us to take tentative steps into the unknown. Learning is simultaneously a leap into the new and strange and also a return to what we already know deeply. Thus, as we move forward on this journey we find that we are returning to our source and remembering ourselves along the way.

My intention in this book is to give the reader a theoretical understanding of Body-Mind Centering, of the principles on which it is based and how it may work in practice, as well as an opportunity to explore the relationship between movement, mind, and feeling states. Through the explorations and exercises at the end of most chapters, I also offer some guidelines for the reader who is interested in experiencing the work firsthand. Some of these exercises may be practiced with another person and can be explored in a process of mutual exchange.

My Own Journey

My own work has been as a dancer, choreographer, dance and movement educator, and practitioner of therapeutic bodywork. Before that, my passions were literature, psychology, and philosophy, ideas, images, and poetry. But some instinct told me I was "all up in the air." I needed to place my feet firmly on the ground and relocate myself clearly in my body. I began to dance as a means to both embody and express who I am. I found I was also on the path of *knowing,* in a new way, that which I am. As I explored ways of making deeper contact with my body, my body was teaching me a new awareness of myself.

In all my studies, I was concerned with the relationships between apparent polarities—mind and body, receptivity and creativity, movement disciplines from both ancient Eastern and modern Western cultures, process and form, healing and art, inward and outward, up and down. These connections were ever-present, each aspect always revolving to its other side, each folding and unfolding into the other.

I was fortunate to find myself, almost by chance it seemed, involved in a wave of new dance activity exploring the interrelationship of mind and body in movement. A form of improvisational dance called "Release Work" was being developed; it used imagery, as physical thought and sensation, to realign the body with the pull of earth's gravity and so create more efficient and healthful patterns of movement. The image guides and informs the action; it is also a source of stimulation for personal creative movement. The activity also affects the feeling state of the body-mind and creates new impressions and images. The continual interplay between thought, sensation, feeling, and action, which is by nature always taking place, is used both creatively and therapeutically in this work.[3]

These ideas and experiences became the inspiration for my future studies and work, leading me into the practice of *t'ai chi ch'uan,* a form of exercise and movement meditation from China that aims at harmonizing body, mind, and spirit. Later, in pursuit of new dance activities, I traveled to America to see and learn more. There, in the autumn of 1979, I met Bonnie Bainbridge Cohen. In the first class I took with her I could hardly understand a thing that was happening! I felt totally confused and disoriented, as if all my knowledge and preconceptions had just been turned upside down. But something had captured my curiosity and enthusiasm—I was excited and knew I must look further. Here was a person who was addressing the questions that most concerned me at that time, and who seemed to have a unique insight into the problems I was grappling with personally. I stayed to train with Bonnie Bainbridge Cohen at the School for Body-Mind Centering, in her remarkable approach to the study of human movement and development. I left with many new insights and even more questions. Since returning home to England I have continued working with these questions and ideas. They are the groundwork of this book and, along with my own students and clients, have been my teachers since leaving the stimulating environment of the school and Bonnie Bainbridge Cohen's own inspiring teaching.

The Origins of Body-Mind Centering

In an interview about her work, Bonnie Bainbridge Cohen states:

> I see the body as being like sand. It's difficult to study the wind, but if you watch the way sand patterns form and disappear and re-emerge, then you can follow the patterns of the wind or, in this case, the mind. . . . Mostly what I observe is the process of mind.[4]

Bonnie Bainbridge Cohen began her career as a dance teacher and as an occupational therapist. Between 1962 and 1972 she worked as a therapist in hospitals and rehabilitation centers, and also studied and taught in New York at the Erick Hawkins School of Dance and at Hunter College. The people coming to her for help with both physical and psychological problems were making remarkable recoveries, but she was unable at that time to articulate what it was that she was seeing and doing to facilitate such healing. It was the desire to understand and communicate to others this natural ability "to perceive, and to help people help themselves"[5] that led her into her research. This work has evolved into the principles and practice of Body-Mind Centering.

This desire to learn took her to train as a neurodevelopmental therapist with the Bobaths in England, working with children with severe brain dysfunction. She also studied neuromuscular reeducation with Barbara Clark and Andre Bernard, *Katsugen Undo* ("the art of training the nervous system") with Haruchi Noguchi in Japan, Laban Movement Analysis and Bartenieff Fundamentals with Irmgard Bartenieff, and dance therapy with Marian Chase. Her studies have been deepened through a wide range of movement and mind practices including yoga, meditation, vocal work, martial arts, and craniosacral therapy. As well as studying with many gifted teachers, she has always learned from the many students—adults, children, and infants—with whom she has worked over the years. Her natural openness, generosity, and humility, together with her tireless curiosity and ability to stay present in the "not-knowing" of the beginner's mind, enable her to learn spontaneously from all situations. In this she shows genuine respect for and interest in each person's gifts and insights, however great or small, as well as compassion and concern for their difficulties.

One of the qualities that has contributed to the uniqueness, in this culture, of her work is Bonnie Bainbridge Cohen's own finely-tuned sensitivity to what is actually happening at very subtle levels of the body and mind, as expressed in stillness or movement. This sensitivity, and an unusual ability to perceive in depth the total pattern of a person's movement and postural expression as well as the flow or obstructedness of the mind which this pattern reflects, are the source and essence of this work. These processes came intuitively to her, she knew them by nature. The material and principles she has been developing and teaching within the framework of the Body-Mind Centering approach, on the other hand, are the result of many years of study and personal research. As Bonnie Bainbridge Cohen herself says, she did not know the theoretical principles and teaching methods. She had to work hard to discover and formulate these in order to develop a language through which she could communicate and teach the essence of her healing work. The material of Body-Mind Centering is the fruit of this research, a language through which the healing relationship between body and mind can be studied and communicated.

On the Nature of Mind

In the Body-Mind Centering approach, we recognize that body and mind have distinct functions; experiencing the body from within, we come to see that they are integrally connected aspects of a greater whole. My own perspective is that both our physical bodies and the thoughts, feelings, images, and so on that are constantly flowing through our minds are but different expressions of that intangible essence that underlies the flow of our individual lives. This essence, which may be called the lifestream or life-force, basic consciousness, inner self, soul, or spirit, manifests itself in constantly changing forms. Body, like mind, is continually in flux, changing from moment to moment in response to the underlying process of which it too is an expression.

Body-Mind Centering language speaks of the "mind" of a particular body system: skeletal, muscular, organ, etc.;[6] or movement pattern. The whole concept of "mind"—its many aspects, levels, and functions, and the different ways people view and talk about it—is a vast and fascinating subject. Some individuals or cultures may use the concept primarily to denote rational thought processes, while others may include aspects such as imagination, feeling, intuition, and so on in their conception of mind. In Buddhist tradition, for example, there is considered to be a primary Mind which is a state of pure awareness of the ultimate nature of reality; and there is a secondary mind in which the ceaseless flow of sensory and cognitive processes obscures realization of the ultimate nature of Mind. Who or what is it that even conceptualizes the notion of "mind," directs attention, or perceives changes in awareness?

A discussion of the nature of mind and the different uses of the term is certainly of great interest, but is beyond my scope or intention here. For the purposes of my description of Body-Mind Centering I will use the term "mind" in quotes when describing a particular body-mind experience. A specific "mind" can be experienced and witnessed when we direct our attention to a particular body system or part of the body, or when we move with a certain focus and identifiable quality. What we experience and observe is a particular quality of awareness, feeling, perception, and attention when we embody a movement pattern or body system; this is the "mind" of that pattern or system, and is an expression of the integrated body-mind.

I will sometimes use the term in its broader, more general

sense to denote the many active-receptive, intellectual, imaginative, feeling, and intuitive functions of mind. In Body–Mind Centering we may focus our mind into the body on a specific area or body system, giving information to the body through visual, verbal, proprioceptive, or kinesthetic means. Such processes help to integrate body and mind by aligning attention, intention, and sensation as they inform the body tissues about movement potential. Ultimately any thought, feeling, or perception moving through our conscious or unconscious awareness affects our body-mind experience, creating a subtle change in the "mind" being expressed at that moment.[7] Thus it is with both the active, receptive, and expressive processes of mind and their integration that we are involved.

I am making a distinction here between mind as it is generally used in Western terminology, as the mental functions of storing and processing information, thinking, reasoning, envisioning, imagining, remembering, directing attention, and so on; and mind as awareness. Let's first look at mind as mental function. Modern Western science and philosophy has tended to divorce the functioning of mental processes from the body, associating them with the brain and viewing the brain as separate from the body. The brain, composed of billions of cells, is of course part of the body. It is intimately linked to all parts of the body through a complex network of nerve fibers and the secretion of hormones and other substances that affect cellular functioning. Psychologically, the differentiation of body and mind out of the infantile experience of psychosomatic unity is an essential developmental process; without it we would remain psychologically immature. But our mentally-oriented culture has created a dualistic split between mind and body and a hierarchy of importance in which mind tends to dominate body to the detriment of both.

Recent research is suggesting what the bodyworker and the intuitive mover know by experience: mental functions, emotions, and bodily processes are not separate, but each influences the others through extremely subtle and complex interactions. In fact, they are inextricably linked by mutually interactive neurochemical processes. Research by Candace Pert, former Chief of Brain Biochemistry at the National Institute of Mental Health (NIMH), has revealed that chemical messengers called neuropeptides and their receptor sites located throughout the brain and other parts of the body form a network of communication linking the brain and the endocrine and immune systems. Dr. Pert's research has important implications for understanding the relationship between thought, emotion, and the body, as well as the nature of body-mind healing, and has led her to the conclusion that

> [N]europeptides and their receptors are a key to understanding how mind and body are interconnected and how emotions can be manifested throughout the body. Indeed, the more we know about neuropeptides, the harder it is to think in the traditional terms of a mind and a body. It makes more and more sense to speak of a single integrated entity, a "body-mind."[8]

Dr. Pert's conclusions lead her to the understanding that consciousness is not located in the head, a common Western assumption, but is projected into different areas of the body. She proposes that

> A mind is composed of information, and it has a physical substrate, which is the body and the brain, and it also has another immaterial substrate that has to do with information flowing around. *Maybe mind is what holds the network together.*[9]

I would like to suggest that the distinction Dr. Pert is making here between mind as information and mind as the immaterial substrata or flow of that information, is similar to the distinction I am making between the mental and cognitive processes of the mind and the function of awareness that can move among, encompass, and pervade all processes and contents of cognition. In Body-Mind Centering we are aiming to integrate the mental and physical aspects of being into a cohesive and spontaneously functioning whole, as well as cultivate awareness in the body. Ultimately, this theoretical distinction dissolves in actual practice.

The Study and Practice of Body-Mind Centering

Body-Mind Centering offers a way to deepen ourselves to the intuitive wisdom of the body and to nurture our innate capacity to heal through awareness and touch. Through it we can explore the very roots of our expression in movement; as we develop awareness of the patterns and qualities of our movement, we come to see how our mind moves or is restricted within the body. Specifically, Body-Mind Centering involves direct experience of anatomical body systems and developmental movement patterns, using techniques of touch and movement repatterning.

In this work we learn through both objective study and subjective experience, attempting to create an integrated balance between the two, rather than keeping them as separate and alien modes of education—the tendency, to a large extent, in modern Western culture. If in this book I tend to place more emphasis on the subjective and experiential aspects of learning, it is because our culture tends to attribute greater value to (relatively) objective and scientific knowledge.[10] I believe that this imbalance needs to be redressed; we can only be enriched by an equal acknowledgment of these two approaches to learning so that they may be integrated into something greater.

Body-Mind Centering has something of great relevance to offer in this respect, as it is based upon observable principles and functions of anatomy, physiology, psychology, and infant development. It is also based on the laws of physics and mechanics as they are expressed through the human body. However, it is just as strongly rooted in the knowledge and wisdom that lie within the subjective depths of each of us. Body-Mind Centering allows our wisdom to speak in its own terms, through its own voice. The language of science or academics becomes a supportive framework of objective knowledge within which the intuitive and "feminine," the body-wise aspect of our nature, is encouraged to unfold. Working in this way, we learn to trust in the value of our own knowing. The theory and principles of this work, in fact, continue to evolve based on our collective experience.

The philosophy of Body-Mind Centering is founded on the understanding that mind and body are integrally connected and mutually interactive expressions of being. Healing or change in the body-mind can be effected by working directly on the body tissues and movement patterns to influence the mind or by working consciously with the mind to positively affect physical conditions. The work is also about learning *how* we learn—how we access information, make transitions from one state to another, and develop the awareness of this process. It is an approach based on direct experience and observation, where the facilitator's own awareness and embodied experience is an essential tool in the educational, therapeutic, or healing work.

Central to the work is the process of awakening awareness at the cellular level to contact the innate intelligence of the body. Awakening cellular awareness awakens "love in the body," as Marion Woodman writes; she goes on to say that "Genuine love . . . permeates every cell of the body."[11] Connecting to another being through touch and presence, resonating with them at this most fundamental level, is at the heart of Body-Mind Centering practice.

In actual practice we work with a continuum of process, from the passive reception of stimulus and sensation given by the practitioner's hands, to the eventual performance, unaided, of a new movement pattern or expression based on the new sensations perceived. Through this approach, the student or client learns to be an active participant in her own process of change and growth. Body-Mind Centering offers a way of bridging the methods of passive bodywork and manipulation of tissues with the more active art of movement reeducation. The body tissues are intelligent. They receive, perceive, and respond to the messages given by the practitioner's hands even before the conscious mind is aware of them. Then, through movement, these new sensations can be organized consciously into healthier patterns of use and in this way are more clearly and consciously established and integrated within the body.

Many of us have been conditioned from an early age to deny the feeling and expressiveness of our bodies. Much of our energy and aliveness is then inhibited, and we also lose access to the knowledge and wisdom that the body holds. Often we feel cut off, disassociated from our physical body, which can lead to sensations of ungroundedness, tension or weakness, discomfort, pain, and lack of real and vital contact with our environment. These sensations will also be reflected in our psychological states and the health of our bodies.

The practice of Body-Mind Centering allows us to gently come back home into the body and to reexperience the harmonious integration of sensation, feeling, mind, and spirit that is ours by nature. We learn to listen to the body through quiet sensing work, our breath, and the use of guided imagery, and to let its wisdom guide us in an exploration of our needs. We learn to trust the body and its intuitive knowledge. Through focused touch and sensitively guided movement, areas of tension and blockage can be released and the core of inner support and strength can be reexperienced. As the body becomes more balanced and integrated, we may experience more alertness and clarity of mind, greater openness and spontaneity of being and perceiving. By freeing habitual holding patterns, we can access and more fully express the creativity within us. This work enhances the body's natural healing ability.

The process of human movement development, from the moment of conception to the mastery of movement on the earth, provides a framework for observation and practice. Development unfolds in a series of stages and movement patterns that reflect the evolution of the species from one-celled organism to humankind. The sequence leads us from our "being" to our "doing" nature—from bonding and grounding to developing a sense of self, reaching out in play, creative action, and relationship. The natural unfolding of the developmental patterns underlies all areas of future learning in the infant, child, and adult. Movement, perceptual, psychological, intellectual, and spiritual growth are all profoundly influenced and supported by these early movement experiences.

The developmental process is both universal and uniquely individual. In Body-Mind Centering practice it is explored and embodied through a series of movement patterns that help us identify areas of personal strength and limitation; movement is repatterned as we allow the "mind" of our learned movement patterns to change. This allows us to move toward greater integration, clarity, and creativity in our expression in the world.

Body-Mind Centering also involves an in-depth and experiential study of all the anatomical systems of the body. The musculoskeletal system and the organs, glands, nervous, and fluid systems each express their own quality of movement, feeling, touch, perception, and attention. These systems reflect aspects of ourselves, and as we embody them through movement and touch these aspects are brought into a more balanced and dynamic relationship with each other. All body systems also provide structural and energetic support for the body; coordinating breath, sensory awareness, and movement with informed touch, we can repattern the flow of energy through any of the body's tissues to facilitate integration and healing.

In the practice of Body-Mind Centering we learn to make direct contact with the different systems of our own body and to initiate movement from them so that each of their qualities becomes available to us as a means of expression. We also learn to contact and recognize the different systems and their movement qualities in another person.

Each body system expresses a different quality in movement and stimulates an identifiable change in feeling, perception, and state of awareness. Similarly any movement done at random, as a purposeful action, or as a specific exercise will express a particular quality of attention, perceptual process, energy, and direction of focus. The action has a specific patterning of initiation, sequence, and completion, which relates to the "mind" of that particular movement. When an individual or a room full of people together are in a certain "mind," this can be recognized clearly; it is often felt as a specific mood or atmosphere.

When we experience directly the anatomical systems and structures of the body, they might be felt to embody inner "characters," the subpersonalities or constellations of energy that coexist within us, acting and interacting with each other in patterns unique to every individual.[12] These patterns may at times remain fixed, or they may change and reorganize themselves into new relationships as our life unfolds.

These constellations of energy, which are embodied in the tissues and structures of the body, express themselves both physically through sensation, posture, movement, and body symptoms, and psychologically through feelings, attitudes, and behavior. In my view, the link between these two levels of expression is not necessarily or purely causal, with bodily activity creating psychological patterns or the psyche determining physical functioning. While they mutually influence each other, they might also be seen as developing together as intimately related expressions of who we are and how we each have journeyed through this life.

Body-Mind Centering is currently being studied, practiced, and applied in their own fields of work by people from a broad range of professions. These include teachers of dance, yoga, martial arts, and other approaches to movement education and therapy; massage and bodywork practitioners; occupational therapists, physical therapists, and psychotherapists; chiropractors, doctors, nurses, caregivers, and health educators; and dancers, singers, musicians, actors, and visual artists. Because of the open nature of the work it can be used as an observational or diagnostic aid in a variety of contexts—a language through which we can perceive and understand our own and our students' or clients' patterns, imbalances, and potential strengths and weaknesses. As an educational, creative, therapeutic, or healing process, Body-Mind Centering also offers an approach that can help us to fulfill that potential. We may address the issues of the student or client at a physical, perceptual, psychological, or imaginative and creative level. This approach can be viewed as both an *art* and a *science* of movement and can open doorways to a rich interchange between our conscious mind and the creative potential of the unconscious.

In applying this work educationally, therapeutically, or transformationally in the creative process, we are looking for a balance and synthesis of energy and expression within each person. Balance is a dynamic process of alternation between different states; as such, it is changing continually. Through the development of awareness we are able to experience this dynamic balance even within the flux of activity, receptivity, and rest.

When individual body systems have been differentiated and brought into awareness, they are reintegrated into the whole to give support and a richer depth and range of experience and expression. We work with the alignment of the physical structure within itself and with gravity; the relationship of the inner environment with the outer; the inner and outer focus of the individual's attention; and the integration of feeling, desire, and intention with attention and action. As the continuity of our awareness develops we will attempt less and less to stop the process of change and hold the moments in a static balance; instead we become more free to relate to what is actually happening in and around us and to dance within the changing moment.

Beginner's Mind

My hope in writing this book is to help bring greater awareness and a sense of wonder to the experience that is your own living body, and insight into the patterns of mind that are made visible and tangible through the body and its movement. The work itself is very much about allowing an open and flexible mind, and if you choose to explore some of the exercises described, I hope you will do so with this same openminded quality.

The nature of learning is that we stand in a state of unknowing in relation to what it is that will be learned. In approaching anything which seems new or a little strange, we need not view it from the attitudes and confines of other systems, traditions, or already established beliefs and concepts, but perhaps "suspend our disbelief" for awhile and enter without too many preconceptions, with the openness of "beginner's mind."

Each time I begin a new phase of my own learning there is an initial period of excitement at the revelation that I am opened up to, the connections that emerge between once fragmented and unknown pieces of knowledge and experience. The threads of this new information grow and form into ever-extending webs and patterns of insight. Yet what I have sensed is still only a possibility, like an image of light and shadows, the vision of a new state of connectedness and wholeness. There is much work to be done before the new knowledge is embodied in me as a living, breathing and loving reality. Then I am truly at the beginning, like a newborn child, about to learn as if for the very first time. I am a beginner yet again, humbled by my ignorance and awkwardness, vulnerable in my inability. And yet I carry somewhere in memory that sense of possibility: a vision of potential, wholeness, and love which acts as a guide for the steps I will take blindly. Silent in the inarticulateness of this knowing, and patient in my trust of it, is the only way I can proceed, if I am willing.

Even as I began this writing, it was from the place of darkness, of not-knowing. It was as if I were about to enter a story whose characters were familiar, like old friends, but their adventure was as yet unknown and untold. Entering into this story, I didn't know exactly how it would unfold, what would be its texture, shape, and ultimately its meaning, for myself and for readers. I took the first step with the faith that the seed would unfold as promised and something of value would emerge. I was challenged to trust in nature's cyclical process and in her mysterious call to come into being and meet the unknown.

This is the beginner's mind. It is the mind, I continually find, with which the Body-Mind Centering process must be approached. I invite you, the reader, to do the same as you explore the work presented in this book. To approach learning from here means letting go of attachment to what is already known and cherished, to seem to forget, to allow the old to die. We are naked, raw, a little naive, emptied of what is unnecessary. In this state of openness there is room for something new to enter. Taking in and letting go is a natural process, like the waxing and waning of the moon, like the expansion and contraction of the breathing of a cell; yet it is often hard for us to allow the dying and the cutting of our attachments without feeling some resistance and confusion.

Each moment must die for the next to come into being. So too our experiences pass continually through the cycle of death, new birth, life, and again death. What we feel to be our difficulties and problems are not these processes of change themselves, but our feelings and attitudes toward them and our reactions to them. Bonnie Bainbridge Cohen has stressed in her teaching that our problems—the places where we feel blocked or the areas of weakness and blindness unique to each of us—are our gifts. If we look deeply enough into these dark places, we will discover something truly meaningful that is of great value to us and of great benefit to others. Wherever we cling to a moment, an experience, a love, there is the question: "Why? What is the nature of this clinging?" This is the beginning of our search for freedom from clinging. We find that at the heart of the problem lies our strength, our reality, and the essence of our own unique being.

The Developmental Process
Underlying Movement

Chapter One

The Ground of Being: Awakening Cellular Awareness

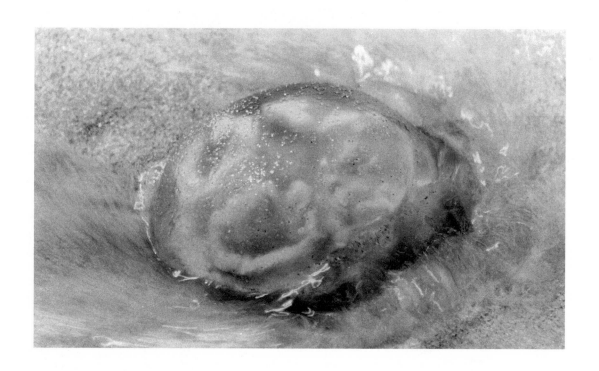

It is through movement that we first learn and establish a foundation for further growth at other levels of our being. The study of human movement development and evolution forms a basis and a framework for Body-Mind Centering work, and so it is here that we now begin.

As with all acts of creation, the life of a human being begins in the place of unknowing. In the darkness of the womb, a cell accepts the seed that awakens the light of consciousness. The sperm enters into the home which will give it new life and form. At this moment of conception the fertile cell becomes a new being, an individual subtly differentiated from the physical body and consciousness of the mother. We begin this current life, this step on our journey, as that single cell with its own unique and individual consciousness.

The consciousness of the cell is perhaps as different from the ordinary consciousness of a mature human as is the latter from the enlightened consciousness of a fully realized being.[1] Nevertheless these and other states of consciousness exist simultaneously within us, many of them beneath or beyond the threshold of our awareness. One way that we can begin to experience these other states of being and perception is through awakening our awareness at the cellular level.

Sentient life on earth, it is believed, began as we did, with one-celled organisms dwelling in the waters of the world, just as the fertilized egg cell dwells in the waters of the womb. The development of the individual human being recapitulates the evolution of the species, from one-celled organism, through fish, amphibian, reptilian, and mammalian forms of life and concomitant levels of consciousness. The cells of the body hold the "memory" of the evolving states of consciousness which were passed through during the individual's early development and also throughout the process of the evolution of the species.

The original single-cell organism, or one-cell, is able to reproduce itself by doubling its chromosomes, which determine the details of sex, structure, function, and appearance of the organism. The pairs of chromosomes then separate to form two identical sets, and the cell divides to form two twin cells.[2] (Fig. 1.1) In the single-celled species there would now be two distinct sister amoebae, each with her own unique life and cellular awareness. This is true also of the fertilized egg cell, which divides and multiplies over and over again. In fact, this process occurs throughout the whole life cycle of a human being, enabling continual growth and renewal.

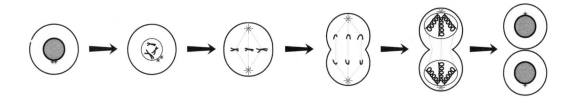

Figure 1.1   The process of cell division.

Each new daughter cell carries and embodies the essential nature of the original one-cell, and has, too, a distinct life of its own. The basic structure of the cell consists of a porous outer membrane which surrounds the cytoplasm, a viscous fluid which contains small particles of nutrients and specialized structures that carry out the essential work of the cell. This work includes the synthesis of proteins, the generation of energy, the division of

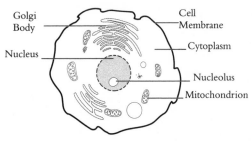

Golgi
Body

Nucleus

Cell
Membrane

Cytoplasm

Nucleolus

Mitochondrion

*Figure 1.2   The basic structure of a cell
(cross section).*

cells, and so on. At the center of the cell is the nucleus, which contains the genes that carry the individual's inheritance and also regulate cellular processes. (Fig. 1.2) A cell can breathe, taking in oxygen and giving out waste gases dissolved in fluid that passes through its porous outer membrane. It takes in and metabolizes nutrients, generates energy, and expels the waste products. Each cell can reproduce itself and contains the potential to perform a number of very specialized functions.

The embryo develops, growing in size and complexity as the cells multiply. Each cell maintains a certain independence and individuality, having as its ground the basic nature and self-awareness inherent in the original cell. It also becomes a small but vital element within the body as a whole, with its own part to play in the interdependence of the two. The cells begin to differentiate and specialization begins. Initially, each cell holds the potential to carry out any one of the many processes necessary to the formation, survival, and reproduction of the entire organism; yet, as if under the guidance of an unseen master plan, this specialization unfolds harmoniously.[3] Some cells develop the

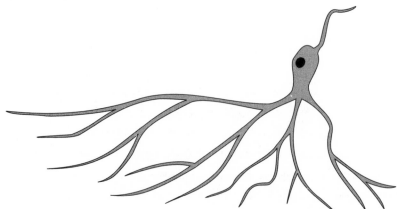

Figure 1.3   Nerve cell.

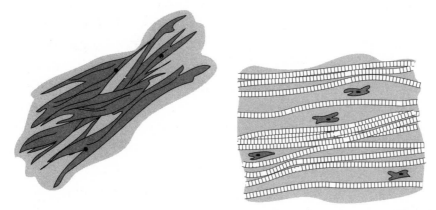

Figure 1.4   Connective tissue.

ability to conduct nerve impulses, and they grow long fibers that carry these messages to other parts of the body; they become part of the highly complex and sophisticated nervous system. (Fig. 1.3) Other cells are able to control the deposition of calcium in the bones and they become part of the skeletal system. Connective tissue cells produce elongated fibers that interweave to create a mesh-like connecting network. (Fig. 1.4) The actual structure of each group of specialized cells both reflects and determines its function, such as the elastic, fibrous bands of muscle cells arranged in parallel bundles. (Fig. 1.5) Many cells will be involved in the special chemical processes of digestion or hormone activity, and so on.

In this way the tissues and organs of the body systems begin to differentiate, and each of these in its turn plays a unique and vital part in the functioning of the whole organism. None is any more or less important than any other for the full potential and health of the individual to emerge. The pattern of life within the individual reflects in microcosm that of life without—from cell to

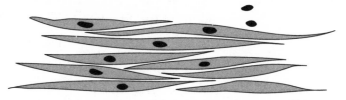

Figure 1.5
Smooth muscle cells.

7

organ to organism, as from individual to organization to society, each small unit plays its part in the greater whole. Clearly, if there is disharmony and disease at the level of the cell, this will affect the proper functioning of the organ concerned and its relationship to other organs and to the body as a whole, which will be thrown into a state of imbalance. Thus any disturbance in cellular functioning is at the root of disorder and sickness in the body. This model can be applied to relationship between the individual and society, the nations of the world and the planet as a whole, and so on, in ever-expanding spheres of inclusiveness. The psychophysical health of each cell is essential to the health and well-being of the whole.

As the biological life of the cell is fundamental to the life of the whole organism, so too is the cellular state of awareness a foundation for normal consciousness, for in one the other has its roots and evolves from out of it. It might be more accurate to call this awareness "preconscious," for it does not have the personal self-consciousness of the adult individual that is, we commonly believe, uniquely human.[4] The cell does not self-reflect or know itself as an individual separate from other individuals and objects. Yet it is responsive to its environment and both influences and is influenced by other cells. To some extent it can determine its own activity. Each cell of the body has its own innate intelligence, its own sense of presence, and its own unique life process.

The "Life" of the Cells

We can make contact with the body on a cellular level through allowing the attention to focus there. Breath, imagery, or the touch of another person also focused on the level of cellular presence and activity can facilitate this change in awareness. (Fig. 1.6) In focusing attention like this we can direct, with our mind, energy to the cells, awakening them to awareness, bringing them to fuller life by enabling them to breathe fully where breath and life-energy

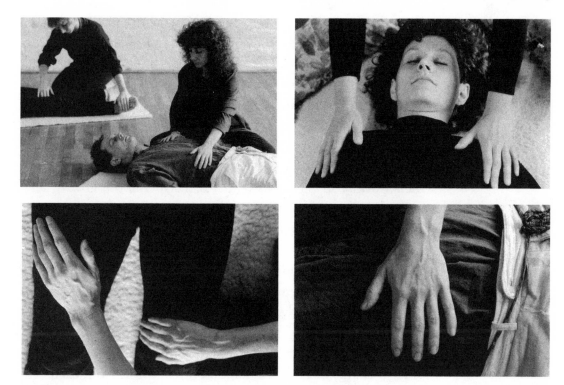

Figure 1.6
Contacting the body
at a cellular level.

may have been restricted. Where the mind moves, there energy is directed and the movement of the body will follow; where it stops, the flow of energy will falter or stop. This principle is fundamental to all disciplines of yoga and healing practice. *T'ai chi* Master Liang says, "When the mind moves, the mind intent is immediately aroused; when the intent is aroused, the *ch'i* will immediately follow. So the heart (mind), the intent, and the *ch'i* are closely connected like a circle."[5]

If the flow of energy to the cells of an area of the body is restricted, the cells lose their vitality and their functioning is diminished. As they gradually die, insufficient energy is available for their complete renewal or replacement. On the other hand, energy may be trapped in a particular part of the body, unable to move on and express. Here, too, the natural flow of energy is blocked: excess energy is locked *into* certain tissues, while denied to others, and may create other kinds of problems. Either situation, and they often occur simultaneously, can over a period of time contribute to states of disharmony or disease. It is now widely acknowledged in both the "alternative" and orthodox branches of medicine that the repression, denial, or holding back of emotional energy is often one of the causes of cancer and possibly one of the most determining causes in the outcome of the disease. Emotional and stress-related factors have been researched in depth by, among others, the Simontons in their work with cancer patients.[6] Rebalancing the flow of vital energy throughout the body is the concern of the many systems of holistic medicine and healing. The art of acupuncture, for example, is based on the theory that there is "a Life Force called Ch'i Energy, and that this Life Force flows within us in a harmonious, balanced way. This harmony and balance is health. If the Life Force is not flowing properly, then there is disharmony and imbalance. This is illness."[7]

When the cells are breathing fully there is a constant supply of fresh energy to and elimination of toxic waste from each cell. This happens through the porous skin or membrane of the cell. With this inflow and outflow each cell continually expands and contracts slightly, in its own rhythm, independent of the rhythm of external respiration taking place through the lungs. This breathing activity of the cells is called "internal respiration," or Cellular Breathing.

Each cell pulses with the movement of its own breathing process, each in its own rhythm; within even the deepest stillness of the body this activity continues ceaselessly. However, we can experience moments of deep peace in which we feel in the cells a stillness that is even beyond this subtle activity. These are moments of integration where every cell simultaneously knows and feels itself and every other cell. Cellular Breathing is the movement pattern which integrates the whole physical body. This state is one that may also be experienced in meditative practices.

It seems to me that this knowing of the cells may be a basis for what we call intuition: a perception, feeling, and recognition of and response, at a cellular level of awareness or intelligence, to that which is too subtle, fundamental, or immediate an experience for our conscious minds to grasp or register. When each cell is present, self-aware, and in potential communication with every other cell of the body, we may perceive information that is normally inaccessible in ordinary states of awareness.

Being and Doing

We can make contact with the cells of specific tissues, fluids, organs, or glands, and experience the "mind" of each distinctly, through directing our attention to that level. Our experience will reflect the functions of that system and will also be colored by our own emotional and perceptual relationship to those functions. The "mind" of the cell, however, is a more neutral, potential, state of "being," basic to the diverse "doing" aspect of the specialization of the body systems' cells.

Before it divides to reproduce itself or specializes in one particular area of activity, the cell is in a state of apparent rest, simply breathing, living, being itself. This is a moment of simplicity, of being present to itself. This moment is not one of complete inactivity, however, for within the boundary of its own membrane the cell is actively engaged in processes of protein synthesis and general metabolism, in preparation for the growth and activity to follow. Any act of creativity requires this period of pregnant rest where many separate elements begin to come together in a state of preconscious awareness. Here, as I write, I will from time to time fall into a state of mental inactivity, non-thinking, non-doing, but sensitive to the ripening of an idea still beyond my conscious reach, beneath the ground of conscious knowing. Then follows a flurry of activity: thinking, articulating, ordering, writing down; then I pause and rest again. Such is the process of the creation of life itself, right from its most basic level, the cell.

Herein lies the therapeutic value of returning at times to the experience of "cellular awareness." The quiet space of mental and physical rest, found as attention is allowed to settle to the simplest ground of physical presence, is essential for the full re-creation of living tissue, as well as creative thought and action. It is the necessary balance and counterpart to the busy activity of the nervous system, which is continually receiving stimulus from the environment, processing this input, conveying messages through the body regarding this information, and directing responses to it. This is exhausting work! Of course we need to rest from it. During sleep much of this activity ceases temporarily, and certain functions of the nervous system can rest and recover. During normal activity, the nerves themselves have a time of rest, known as the "refactory period," between the conduction of one nerve impulse and the next down any particular axon. But because of the extreme stresses and complexities of our modern society, characterized by a constant pressure to be "doing," we may not allow ourselves to go into these periods of rest fully. An agitated and anxious mind will not enable full rest and recovery, so that stress perpetuates stress, instead of creative activity flowing naturally out of quiet rest.

In states of deepest unconsciousness, in "coma," we see the most extreme example of the body-mind functioning at a cellular level. Here most of the essential life-supporting functions of the nervous system can no longer operate without the use of critical life-support systems. Consciousness slips into a darkness close to death. After severe injury to the brain, for example, the traumatized nervous system needs to rest very deeply, allowing the body to exist for a while at the level of cellular activity and awareness, supported by artificial means, while all available energy resources go into the healing and reorganization of the damaged parts. We might compare this situation to that of the fetus whose life is sustained by the support systems of the mother's body as it develops its own readiness for independent life.

In a television documentary I saw, the journey of a young girl through the mysterious darkness of coma after a serious accident was recorded. I was moved to see how the loving contact of her mother and devoted nurses brought her back to life. As she lay curled up in fetal position, apparently unaware, her mother touched, held, and spoke to her as to a newborn child. Eventually she began to respond. Over long months and years, like an infant, the young girl gradually learned to walk, talk, even dance— to live again. Hers was a story of great hope, faith, and courage, intense pain and frustration, and also of joy. This is the story of all life renewing itself.

Let us remember that each system, including the nervous system, is made up of millions of cells. We can contact either the "mind" of the specific system, for example, endocrine, nervous, or skeletal, or the "mind" of the cells in general of each and every body system together. In going to the cellular level of awareness, we simultaneously contact the "mind" of every cell of the body in its essential state of neutral potentiality, simplicity, and rest. This is a more basic state of awareness, before differentiation into systems; in a balanced state of being, activity will be grounded in such awareness, emerging from it spontaneously. The activity of the cells of every system needs this grounding in simple presence.

When we focus on the presence and breathing of the cells, experiencing the "mind" of the cells at rest, we do not lose consciousness altogether, although the state of consciousness does change. Nor is the nervous system going to cease its activity when we focus on its cellular presence—of course this only happens when we die! But we can allow our wakeful mind to come to rest in the essential nature of the one-cell, in the particular quality of awareness that focusing on the presence of the cells evokes. In doing this, we should be mindful that this experience of restful being is at the source of all activity and is not a state in which to attempt to remain forever. To try to hold on to such a state denies the natural cycles of change and would in a way be a negation of life, for "doing" is as necessary to "being" as rest is necessary to activity. Our aim is to look to the artful balance of these two aspects of cellular activity in all levels of the creative process. If our "being" and "doing" become divorced from one another, if one mode habitually dominates our lifestyle and expression, we will drive ourselves eventually to a state either of nervous exhaustion or inertia and apathy.

The nervous system carries the impulse toward life and fulfillment through action, but if it is separated from its origins in the nature of the cell, if we try to function solely from this "doing" aspect of the nervous system, we lose the roots of the meaning of our life in the endless activity of ever-increasing desire for more. On the other hand, a tendency to live from the "mind" of the cell alone, in the eternal, timeless present of being at-one, can result in a withdrawal from life and from the expression of our being through creative action and relationship if we have not understood the real meaning of "not-doing." When we can dance between our "being" and "doing" selves and create a balance of harmonious interaction between the two, however, then each becomes a support for the other. In the expression of the opposite, each is able to recuperate and find a way towards its own fulfillment. It is the law of the cycle of change that out of one extreme arises its opposite—even as we pass through the heart of night we are approaching dawn. So too with the processes of rest and activity. If we go fully into the heart of one, the other will naturally follow. But if we don't go fully there will not be a full return.

The state of being of the cell may be experienced as a feeling of rest, peace, and simplicity. But in it there can also be a feeling of great power and omnipotence; the original cell is the center of its universe—it *is* its universe. The form of the cell is the sphere, which expresses a quality of wholeness and completeness.

As the multiplying cells grow through the embryonic and early fetal phases, this feeling of omnipotent at-one-ness predominates as the primary mode of primitive awareness. The fetus experiences no boundary between itself and the whole world in which it exists, being integrally connected as it is through the umbilical cord to that world that sustains and nourishes it. Everything it needs for its existence is there, always. It need do nothing to receive the nourishment and holding it requires to live and grow. Physically it is at one with its mother's body, and through this unity it experiences itself as at one with the universe in eternity. There is as yet no sense of the progression of time. This is the state of awareness, the "mind" of the cells that are growing into a human being.

Out of this a sense of the power of being is established, the ground from which self-consciousness will later evolve. This is a kind of bonding to oneself, fundamental to the bonding with earth and mother and later to community, that can happen when a sense of self separate from other begins to emerge during and after birth. It is a feeling of at-one-ness and omnipotence to which we may return throughout life. But if we are to evolve through the life cycle and not remain permanently in a womb-like state of "preconscious unity," then we must first travel forward through the steps of differentiation, separation, and self-consciousness that we understand as the process of physical and psychological development of the infant and child.[8]

The Roots of Learning and Perception

The growth every child and adult goes through is not so much a linear as a spirallic process. At the beginning of each new phase of development, each new turn of the spiral, we will consciously or unconsciously again pass through the shadows of the first phases of our life. Reflections or reenactments of patterns first experienced in the womb or at birth can be observed in the movement development of a young child, and this sets the foundation for future stages of perceptual, emotional, social, intellectual, and spiritual growth.

The first learning takes place in the womb, primarily through physical sensation, and at this stage body and psyche are not yet differentiated. Here the wisdom of the body begins to develop, first through the simple and direct knowing of the cell, then through the complex sensitivity of the nervous system.

Even in the watery environment of the womb, the cells are learning about their place on earth through the sensations of the pull of earth's gravity acting upon them. As they float in their timeless world they constantly receive information about themselves and their relationship to the world they inhabit from the mother's movements and the great slow rhythms of the earth itself. At this cellular level we first learn to surrender to the great force of gravity, to let ourselves be supported by it and respond to it. "On earth, where we dwell, our mass or body has weight— each infinite particle of mass being acted upon by its own specific gravitational line or force."[9]

When going through the cycles of growth as children and adults, we fear the loss of connection to a supportive holding environment. As we make transitions to higher levels of the spiral, it is through bonding at the cellular level that we can reestablish our connection to the earth.

Along with the sensations of movement—its own and those of its surroundings—the perception of touch is the primary way in which the cell learns about itself and the environment with which it comes into contact. The outer membrane of the cell touches the fluid in which it floats, which in turn touches other cells around it. All are minutely moving and breathing, giving and receiving stimulus to each other, and upon this contact a rudimentary sense of self-knowledge is founded. Deane Juhan writes:

> Touch is the chronological and psychological Mother of the Senses. In the evolution of sensation, it was undoubtedly the first to come into being. It is, for instance, rather well developed in the ancient single cell amoebae. All the other special senses are actually exquisite sensitizations of particular neural cells to particular kinds of touch: compressions of air upon the ear drums,

chemicals on the nasal membrane and taste buds, photons on the retina. . . . Touch, more than any other mode of sensation, defines for us our sense of reality.[10]

Through the perception of touch and movement, the cells of the growing fetus are beginning to learn about their own presence and activity and the variables of the world they inhabit. The qualitative experience of this early learning will have a great influence upon the way we experience learning and growth in future phases of our development.

The quality of holding and stimulation experienced in the womb will affect our later passage through transitions in the life cycle, and will be reflected in individual patterns of response and reaction to them. However, if we can allow ourselves to settle into the level of cellular awareness and reexperience ourselves in this way, through cellular contact that is nurturing and holding, we can begin to create a more positive experience of this early learning phase upon which new patterns of response can then be built. Out of the experience of what we call "cellular holding," we can, as children or adults, be guided again through early learning processes that may have been first experienced as unsafe, traumatic, or incomplete. With positive support, love, and guidance we can reexperience these stages and transitions in their original wholeness, as they would be when all conditions allow nature to do her work unhindered. These are moments full of trust, challenge, excitement, joy, and the sense of self-worth that comes from knowing something new has been mastered.

Memories of very early experiences are "stored" in the body as energetic blocks and physical sensations, and awareness of these may be evoked at moments in later life in which some associative connection with the original experience arises. This usually happens below the threshold of conscious awareness, through a bodily reaction, symptom, or attitude that in turn affects other areas of behavior. Through bodywork and movement therapy, new sensations can be experienced and different patterns of response explored. By giving the appropriate support and stimulation for a new pattern of sensations to be integrated, we then have a choice between the new and the old way of experience. Once a new pattern of health and wholeness has been experienced and recognized, the body-mind will tend naturally toward choosing this in preference to a less efficient and harmonious one, providing this pattern is given enough support for the new sensations to be integrated into the nervous system. In this way we have the potential for a stronger and more secure foundation for future growth. Like trees, we can seek far down in the earth to grow deeper roots, reaching toward the source of the earth's energy.

Exploration: Cellular Breathing

The process described below will help you to make contact with the restful and recuperative state of cellular awareness and access a feeling of being truly and powerfully present in your body. Cellular awareness also provides a basis for the explorations suggested in the following chapters, opening us to the receptive and intuitive processes necessary to this work.

It may be useful to make a tape recording for yourself of the following suggestions, adapting the ideas to suit your own needs and inclinations. Leave enough time after each stage so that you can experience it as fully as you wish. Even better, ask a friend to talk you through the exercise. The directions can also be made in the first person as an affirmation, such as: "I feel my body making contact with the earth." Make sure that you do the exercise in a warm and comfortable place with sufficient space in which to move around, if possible.

Begin by lying on your back on the floor. (If you wish, you can use small cushions under your head and knees for greater comfort.) Close your eyes.

Feel the places where your body makes contact with the

ground; sense the floor spreading out around you, in all directions, supporting your weight; imagine the foundations of the building reaching deep into the heart of the earth, too. If the weather is warm, it can be particularly enjoyable to do this exercise outdoors. Otherwise you might like to imagine you are lying on the earth, perhaps on grass or a sandy beach. Feel the warmth of the sun and the rich life of the earth directly beneath you. Allow yourself to take in the pleasure of the sounds and smells in your environment.

Let your body soften and spread to meet the ground that is holding you. Feel your skin opening to receive the touch of earth and air.

Now observe the movement of your breath flowing gently in and out, connecting your inner space with the surrounding space. Notice where in your body the rising and falling rhythm is deepest; as you relax more this movement may change.

Can you feel this filling and emptying motion spreading through your whole body, into the chest, the abdomen, and right down into the pelvic area? Allow the breath to move you. Can you imagine it going into your arms and legs, as far as your fingers and toes, and up into your face and head? (As your attention travels like this around your body, there may be places where it is hard to feel or imagine the breath going, areas that seem dark and difficult to contact. Don't force the breath; just observe.)

Become aware that your body is made up of billions of tiny living cells. Each cell is different, but each has the same basic structure: a nucleus at the center surrounded by cytoplasm, which consists of 70 to 80 percent water and molecules of various kinds, and a semipermeable membrane that envelops the cell and forms its outer boundary. Through these membranes the cells are breathing, minutely expanding and contracting, out from and in towards the center, in a pulse of life taking place throughout the whole body.

As you listen, the breathing of the cells may give rise to

the perception of a subtle pulsing, vibrating, or tingling, a sensation of heat, or perhaps an undulating rhythm of movement throughout your whole body; there may also arise a sense of deep stillness and peace. Be open to perceiving a sensation of the cells breathing, to feeling this rhythm. This is not the rhythm of the breath coming in and out of the lungs, nor the throbbing of the heart that you may also be able to feel; go deeper.

Again you can let your attention travel into areas where before it was harder to feel your breath or make contact and allow the cells there to breathe fully.

Feel the body as a whole, every part alive with this very subtle pulsation. Hold lightly in your awareness the knowledge that your body is one connected entity that at the same time consists of billions of tiny individual breathing cells.

Allow each cell to release its weight into the support of the earth. Let your mind focus on any areas where you feel the cells drawing away from gravity; let them relax and breathe freely. Allow yourself to dream.

Feel again the earth spreading out to support you, carrying you through her own great movements through space.

When you feel ready, roll onto one side gently and with as little effort a possible. In this new position feel the pull of gravity acting on each tiny individual cell. Relax and enjoy the sensations.

Roll again onto the other side, and rest. This is a moment of eternity.

Then roll softly onto your stomach. Feel how the roundness of the earth holds you. Open out to embrace her in return, and experience both holding and being held.

For the next part of the exploration you may wish to play some music of your choice. Now let your body begin to make whatever small movements it wishes. Your body knows what it needs to do, so allow it to lead you into moving to explore its possibilities and its boundaries. Be aware of the contact you make through your skin with the ground and the air. At some point, when it feels right, open your eyes and take in the sights around you as you move. Feel the floor, the walls, the ceiling of the room, and the objects and people in it all supporting your movement. Feel that you are also supporting them. Take as much time as you need to eventually bring yourself up to a sitting or standing position.

When you feel truly present in the room again, acknowledge the ending of the exercise in your own way.

It can be helpful to make some notes or a drawing of your experiences, or share them with a friend. This can bring you gently back into ordinary awareness again, and can also help to make conscious and integrate any important feelings or insights you may have had.

Chapter Two

The Pattern Unfolds:
Movement Development in Utero

W e have been considering the cell, the basic unit of organic and sentient life on earth, as a seed containing the potential for both the development of more complex and differentiated physical forms and the emergence of higher levels of consciousness. Enfolded within the cell's very structure are the patterns—not yet manifest, but implicit in their very nature—of all future stages of the organism's physical development. The whole is contained within the seed right from its conception.

Although there are infinite variations in the ways these patterns will be expressed by the individual, the essential nature of each stage is universal to the species. The unfolding process follows a sequence that can be observed in the evolution of all animate life, up to the level reached by each species on the evolutionary scale. The process of evolution, which begins with the cell, is a series of transitions through increasingly sophisticated levels of form and function. And just as the cell has embedded within it this potential for physical evolution, so too does it carry the potential for the development of consciousness, which we have called, at this embryonic stage, cellular awareness or cellular "mind." Each subsequent stage of physical development manifests and expresses an increasingly higher degree of perception, awareness, and consciousness—in essence, the "mind" of that stage of evolution.

The development of consciousness parallels that of physical form and function in the evolution of the species and also throughout the growth of the individual human child from cell to embryo, fetus, and infant. In observing the development of movement in the fetus and the young child, we can see an unfolding pattern that is reflected in all levels of development. The very process of learning, of making transitions from one place to another, and of undergoing transformations in fundamental nature, is made visible and tangible. The journey the young child takes through each turn of the spiral of its physical growth can serve as a model for its future steps in other areas of development, and contained within the process of movement development are the potential and foundation for these steps. Ken Wilber writes:

> The fetus "possesses" the ground-unconscious; in essence, it is all the deep structures existing as potentials ready to emerge, via remembrance, at some future point. All the deep structures given to a collective humanity—pertaining to every level of consciousness from the body, to mind, to soul, to spirit, gross, subtle, and causal—are enfolded or enwrapped in the ground-unconscious.[1]

His discussion goes far beyond the realms of physical human development, through psychological, mental, and spiritual levels of consciousness. But let us look further at the physical development, seeing it as a stage within the greater context of the fully evolving human being. It is both part of and a necessary foundation for the development of all levels of being; it also implicitly contains the whole. The body and its movements provide a physical basis for consciousness and are the medium through which this can be embodied and expressed in human activity.

The First Movements

Each cell of the human embryo contains within its genes all of the information necessary to the creation of human form, together with the individual variations within that form and functioning. As the cells divide and differentiate into systems, it is the DNA molecules which direct the unfolding of the plan, a kind of blueprint of the series of stages through which the child-to-be will progress. Influenced by the communications of the nervous system, together with the activity of hormones, this blueprint is embodied in form. The body systems develop the structures and functions appropriate to each phase of growth, are stimulated to make changes when necessary, and are integrated into an ever-changing yet coherently whole organism.

After the cell, the second recognizable form to emerge is that of another water-dwelling creature, the starfish. The physical form of the starfish is organized around a center from which its five limbs radiate. The starfish's mouth and "brain," the location of coordination of nervous activity, are at the center, while at the end of each limb are sensory receptors and a pigmented light sensitive spot that serves the function of an eye. In human beings, this structure will later be reflected in the sensitivity, through touch and vibration, of the hands, feet, tail of the spine, and special senses of the head. The starfish's limbs are all used equally in locomotion and the grasping of food. This equal use is called radial symmetry. (Fig. 2.1) Food is ingested through the mouth at its center, though in the common starfish Asterias, the stomach actually extrudes from the mouth to engulf its prey directly.

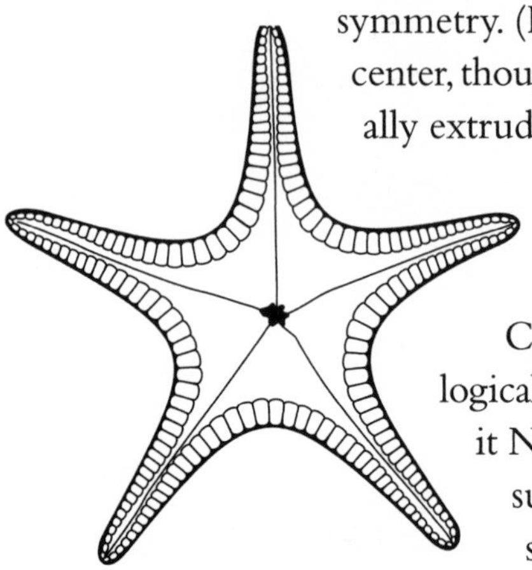

Figure 2.1
The starfish—the limbs radiate from the center.

The radial form of the starfish is reflected in the human fetus, which is connected by the umbilical cord to its mother's body. Bonnie Bainbridge Cohen has identified this as the second basic neurological pattern within the developmental sequence, naming it Navel Radiation. From the umbilical cord the fetus is suspended, floating in the fluid within the amniotic sac; through the umbilicus all nourishment and energy needed for its growth are received, and waste products are eliminated. Like the starfish, its "mouth" is located at the navel center from where the life energy flows to the whole organism. (Fig. 2.2) From the center of the human embryo and fetus six limbs develop. The neck and head are first to develop and most

pronounced; the closing of the neural tube first happens in the
neck, then proceeds upwards to the head and down to the tail;
the neck is thus an important neurological area in terms
of movement development and perception. Then, a
very distinct tail and the two upper limbs develop,
followed by the two lower limbs at a slightly slower
rate of growth. The head of the fetus, with the
mouth dominant in early development, evolves
from the centrally located mouth of the starfish
which elongates out of the body to grasp its food;
the other five limbs reflect the five limbs proper
of the starfish. This growth order of the limbs will
recur later on as the infant develops the neuromuscular coordination to use them purposefully in locomotion during its first year after birth.

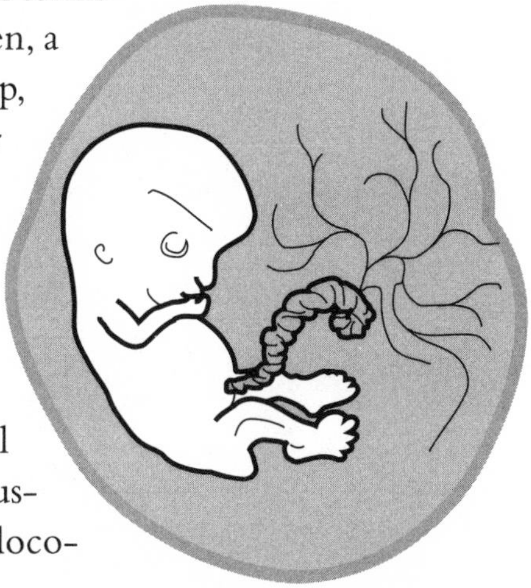

Figure 2.2
The fetus (eight weeks
old)—the center of
organization is at the
navel.

It might at first seem that the fetus would not necessarily
need to move at all in order to ensure its biological survival. It is
passively fed all the nourishment it needs continuously and is protected, warm, and safe within the womb. However, movement is
in fact essential for the future physical, sensory, perceptual, psychological, and mental development of the child. Through movement the fetus' nervous system develops, awareness of itself and
its environment begins to emerge, and a foundation for future
learning and modes of interaction and response is established.
The health and future realization of the individual's full potential depends, at least in part, on the experience of itself as a moving being in this early stage of life.

The first nerves of the body to myelinate (a process by which
nerve fibers are sheathed in a fatty insulating covering, greatly increasing their conductive ability) are the vestibular nerves. They
register information about the movement of the fetus and its
environment: mother and earth. As the fetus moves and is moved
within the mother's body, sensory information from the vestibular nerves is processed within the central nervous system. Per-

ception of these changes stimulates more movement or a change in movement, which again elicits new sensory information. This sensory-motor feedback mechanism of the nervous system provides a basis for developing awareness of self and differentiating self from other. The fact that the vestibular nerves are the first to myelinate is an indication of their importance for growth and survival. It is also of great significance that motor nerves myelinate before sensory nerves. We move first, then we receive feedback through sensation about that movement. Bonnie Bainbridge Cohen states:

> [W]e learn first through the perception of movement. Not only is movement a perception, but as the first perception of learning, it plays an important role in establishing a baseline for our concept or process of perceiving. This original process of perception then becomes incorporated into the development of the other perceptions.[2]

Movement is registered not only through the vestibular nerves of the inner ear, but also through proprioceptive and kinesthetic nerves located in bones, joints, muscles, fascia, and ligaments throughout the body, and through interoceptive nerves in the organs, glands, vessels, and nerves. Movement is also registered by each cell of the body. Information is received from each of these sources about where the body is in relation to gravity, space, and time, and what the quality of rest or activity is, all of which is vital both to the evolving consciousness of self and other and to the dance that can happen between them as this consciousness grows.

Touch also plays an essential part in this process. As the fetus moves within the womb, the mother's body, which comprises its universe, moves around it. Through this movement, the fetus makes contact through its skin with the fluid and lining of the womb, with the mother's organs through the walls of the womb, and with its own body parts. Touch pressure, rhythm, and vibration are sensed, and so the fetus receives information from the environment in response to its own movement. This will stimulate further activity on its part, and so on. It is through this very process of sensory-motor action and interaction that connections between nerve cells are made, and so the potential for learning, for more fully experiencing life, is nurtured and encouraged to emerge. In later life the activities of purposeful movement, speech, and thought as we interact through ever subtler movement forms with our universe will also give this feedback and will produce response. Each future pattern of intention, activity, and response has the roots of its design embedded in this early learning experience.

The Navel Radiation Pattern

Movement in utero, then, serves a very important function but is not yet based in conscious intent. Intention-directed movement will only emerge later, during and after birth, as it becomes necessary. Function unfolds with the progress of evolutionary necessity and also determines the form. There is, however, a pattern and unconscious purpose to the movement of the fetus, which could be described as pre-intentional, a shadow or an implicit design of patterns that will later evolve in more explicit form.

By eight weeks of age the embryo has grown into a recognizably human form, a tiny fetus with all of its organs and limbs already developed. From then on it will grow in size, and through the stimulation of movement, touch, and vibration, the potential of the nervous system will begin to unfold. The dominant pattern of movement in utero we call Navel Radiation; this is the same pattern of radial symmetry that is found in the starfish. This pattern can also be seen quite clearly and vigorously in the newborn infant. As it snuggles in, throws back its head, or thrusts out with an arm or leg, these actions appear to originate not locally, in the muscles of the limbs, but from the navel center of the infant's body. At this

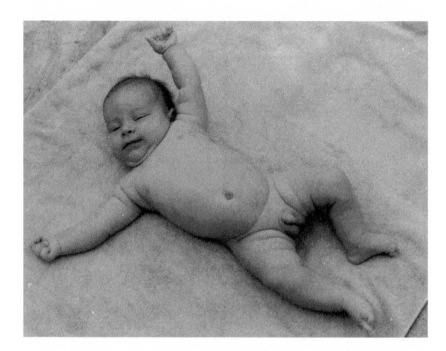

Figure 2.3   This two-month-old baby still shows the Navel Radiation pattern.

earliest stage of pre- and postnatal development, movement is organized around the navel center; from here it radiates through all the six limbs of the head, tail, arms, and legs. (Fig. 2.3)

Through this process all of the limbs are clearly differentiated and then reintegrated into an articulate whole-body pattern. Each body part will learn that it can initiate movement independently of the other parts, but is at the same time connected and related to them through the navel, affecting and responding to the whole. This process of differentiation and integration will be seen again at each level of development of movement and consciousness and is also basic to all methods of education and therapy, both physical and psychological. As we differentiate, we can dis-identify from the part, then reintegrate it at a new level of wholeness and awareness. This enables us to relate with a higher degree of consciousness and skill to the individual parts of ourselves and to the environment without losing the integrity of the whole.

Each limb experiences itself moving separately and also in communication with each of the other limbs. At this early stage of development, another kind of movement parallels in development that which emanates from the navel center. The organization of a number of primitive reflexes is taking place within the nervous system.[3] These reflexive movement patterns are coordinated at the level of the spinal cord or lower (primitive) areas of the brain and therefore do not come under the conscious volitional control of the higher areas of the brain. They are particular coordinations of activity elicited by specific stimulation.

Primitive reflexes emerge at specific times in pre- and postnatal development. They are stimulated by touch or pressure to particular areas of the body, passive movements of the head, torso or limbs, changes of position, changes in relation to gravity, or sudden and unexpected sounds, movements, etc. The infant responds to the stimulus by moving toward it or drawing away; these responses support the potential for bonding and defending, both of which are necessary for healthy movement development and also personal survival. Reflexive movement responses also serve the vital function of developing balanced muscular tone and coordination throughout the body, in preparation for eventual crawling, standing, walking, and so on. The more complex righting reactions and equilibrium responses develop subsequent to the integration of a number of primitive reflexes. Righting reactions establish the axis between body and head and enable the infant to maintain an upright head position in relation to gravity. Equilibrium responses allow the child to maintain balance or protect itself when about to fall; these responses are present throughout life.

The development and integration of these automatically controlled movement responses is essential to the normal development of movement. They underlie the basic developmental movement patterns that will emerge from birth onwards; a number of specific reflexes need to be integrated in order for each of

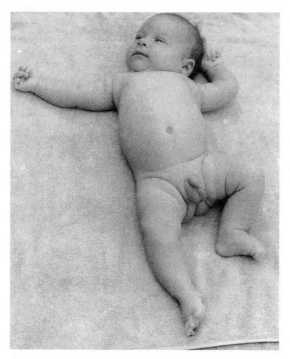

Figure 2.4   The Assymetric Tonic [Neck] Reflex underlies hand-eye coordination and Homolateral crawling; this reflex is one of several which must be integrated in order for the Homolateral crawling patterns to emerge.

the developmental patterns to be mastered. (Fig. 2.4) The newborn infant will tend to respond to specific stimulation through the appropriate reflexive actions and reactions. Reflexes are integrated when the infant is freed from these predetermined responses and can react to such stimulation with greater choice. The reflexive movement patterns, however, continue to act as an unconscious support for more complex and volitional movement. Their underlying presence continues to give clarity, ease, strength, and a graceful quality to movement throughout life. To reawaken these qualities, underactive reflexes can be stimulated even in the adult. Work with the reflexes is used traditionally in the practice of physical therapy and neurodevelopmental therapy, where it has particular relevance to the brain-injured child or adult. Such work is not normally considered to be of value, or even desirable, for the "normally functioning" person. What we are now seeing is that the actively supporting presence of the reflexes is necessary for the full expression of potential and aliveness in movement.

The developmental patterns in turn underlie all human movement possibilities, from the positions we adopt while sleeping to the way we hold a pen or perform the most intricate dance movements. As the developmental patterns are gradually mastered, the child is freed from the predetermined responses of the reflexes and can act upon its environment from a place of greater choice and volitional control. However, the emergence of the primitive reflexes in utero and immediately after birth is an essential stage. It allows a transition from the arbitrary and preintentional activity seen in the Navel Radiation pattern to purposeful and willed action.

Movement happens through the cells of the body. When each cell is breathing freely it is in continual communication with every other cell, each responding to and influencing all the others. This happens through their subtle movements, the pulse of internal respiration, and chemical processes taking place within them. For the limbs to integrate into the navel and move freely from it, the cells need to be alive, breathing, and responsive. If cellular breathing is inhibited anywhere in the body, then the related limb will not be fully integrated into the pattern of navel radiation, and this will result in weakness or difficulty in the development of future movement patterns and perceptual responses after birth. Often developmental problems in children or adults can be traced back to an inhibition in cellular breathing and a lack of integration of the limbs into the navel at this stage.

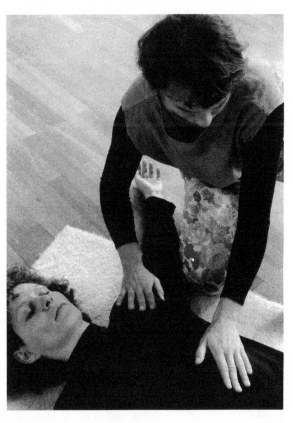

The connection of the navel center to the extremities is made through all of the body tissues at a cellular level; integration of the limbs happens through bones, muscles, ligaments, organs, glands, nerves, fluids, and connective tissues. Each area of the body is in direct relationship with a particular limb, and all tissues and organs within the area are involved in and give support to the movement of that limb. The blocking of energy flow and connection through a limb can be happening in the cells of any organ or tissue layer or in the fluids themselves. In Body-Mind Centering work we seek to identify the level at which flow and integration are inhibited, and by awakening cellular awareness in

*Figure 2.5
Integration of the limbs
into the center through
awakening cellular
awareness in the tissues.*

the tissues of that area or body system, connection can be made or remade and the flow of energy and movement guided back into its natural course. (Fig. 2.5) The developmental patterns give a framework for this, the natural unfolding of movement potential, but before we move on to more of the explicit forms embodied by the developing infant, we can explore the primary underlying connections through the Navel Radiation pattern.

Exploration: Navel Radiation Pattern

Through exploring the process outlined below you may be able to recognize areas in your own body where the integration of the limbs into the center is blocked. This might be experienced as a difficulty in focusing awareness there, or in imagining the connection of the flow of breath through the limb. Practice with this exploration can facilitate a sense of connectedness and integration, give a deep experience of inner support for movement, and allow the whole body to relax and breath more fully. As with all the exercises presented in this book, your own experience will be unique; be open to the sensations, perceptions, and insights that may emerge. Give yourself enough time to experience each stage as fully as you need to at that moment.

Begin by finding a comfortable place to lie down, preferably on your back. Close your eyes and spend some time focusing on Cellular Breathing, as in the previous exercise. (These two exercises can be explored together, to help you experience their interrelationship.)

Imagine your breath entering through the navel, filling the middle of the body, front to back and side to side, and radiating from there to all parts of the body on the inhalation; as you exhale and empty, imagine that the breath flows out through the navel again. Keep imagining this movement of the breath until you can begin to feel the flow of energy, carried by the breath, through each limb—from the navel to the fingers, toes, top of the head, and tail of the spine, filling as you inhale—then returning back to the center again, emptying out as you exhale. Stay with this until you can feel the sensation of this subtle movement of breath spreading equally through all six limbs. Focus on allowing the cells to breathe wherever you experience a lack of connection.

Let your concentration be light, delicate, alert, and fluid. Concentrating too hard to create the image and sensation of the breath moving in this way will only cause unnecessary tension.

Allow the breath to move you. Begin with small internal movements through each limb, extending out and compressing or folding back in to the center with the flow and rhythm of the

Figure 2.6
Exploring the Navel
Radiation pattern in
improvised movement.

35

breath; try to maintain the feeling of connection of the limb to the navel, so that the center and the extremities become aware of each other through and throughout the movements.

You can then explore connections between two or more limbs simultaneously, feeling that they relate to each other and know each other through the navel. Through the center a dialogue is taking place between the head and tail of the spine; the two arms all the way to the fingertips; the two legs and feet; the right arm and the right leg; the left arm and leg; then the right arm with the left leg, and the left arm with the right leg. There are also connections between the hand and face, the foot and tail of the spine, the tail with both hands together, and the head with both feet. Explore in movement all of the possible connections and relationships, allowing your body to find its own expression of this as the limbs move together and apart, touching each other and the space around you in a dance of separating and rejoining.

Let yourself enjoy the sensations of moving and being moved by the flowing of your breath. Allow the breath and these internal connections to actually support your movement; the feeling is that of moving in water, or as if you were suspended within a large bubble. With your limbs gently explore the boundaries of this sphere and the surface of your own body.

As movement becomes a little more active, allow your whole body to roll, stretch, curl up, wriggle, wind, and unwind. There are infinite possibilities, and you have all the time you need to explore them. (Fig. 2.6.)

Feel the connections through the body from the center to each extremity as you move on your side and stomach, too. Allow these connections to support you as you explore the dialogue between the limbs and the center in more active movements. Include in your awareness the sensation of your skin touching the floor, the air, and other parts of your own body.

Begin to feel the floor with your hands, feet, head, and tail, allowing the earth to support you through them. As you release your weight into the ground, feel how it responds by supporting and pushing back, right through your center and beyond. As your movement becomes even more active, open your eyes and receive your environment through them and feel how it, too, supports you. Allow yourself to become present once again to the room. (Again, you might use music for this last part, opening your hearing to the sound and letting it, too, move and support you.)

You may wish to come gradually up onto your feet, using the internal connections of the Navel Radiation pattern for support as you move from the floor through changing levels—rolling, sitting, kneeling, squatting, hands and knees, and so on. From here you can explore the relationships of limbs and center in improvised dance movements.

Finally come to stillness to acknowledge the end of the exploration, maintaining awareness of your center and six extremities and the sense of connectedness throughout your whole body.

Again, it is useful to make some notes or a drawing to help you ground and integrate your experiences.

If you completed all the stages of this exploration, you probably went through most of the basic developmental sequence in your own individual way. Your experience in this and the previous Cellular Breathing exercise can reveal much about your own developmental process. You can explore in this way as often as feels beneficial to you. If you found the Cellular Breathing and Navel Radiation exercises difficult in any way, however, or were unable to fully "come out" and feel yourself present at the end, it would be advisable to read the next chapters on the developmental process before exploring these exercises further. This will help you learn how and where you need to direct your focus in order to use each experience in the most beneficial and creative way.

The "Mind" of the Navel Radiation Pattern

Each person's experience of Navel Radiation will be unique and uniquely valid. We can attempt, though, to describe in general terms the essential state of awareness and perception of each pattern by looking to the form and quality of movement manifested, as well as the purpose of that phase of development.

Conception could be said to be the first stage of separation in an individual's life: the egg cell, once fertilized, becomes a separate living entity with its own consciousness, distinct in a very primitive fashion from that of the cells of the mother's body. Also, in the union of the two cells from mother and father is the first intimation of differentiation and duality. Both unity and duality are already contained, in embryonic form, within the consciousness of the seed of a new life. However, this is as yet beneath the threshold of conscious awareness, and it is not until birth that the transition to consciousness of being a separate self really begins.

During life in utero the fetus experiences itself primarily in the state of omnipotent "at-one-ness" with its universe; the fetus is the universe itself. As with the one-cell, the developing fetus lives in a world of suspended time because time is not yet known to it. Body and self are experienced as inseparable, consciousness and feeling as physical sensation. If the womb is perceived as "friendly" and nurturing, early sensory experiences will be imbued with pleasurable qualities; such experiences form a foundation for the development of self-love and a basic sense of trust in a benevolent universe. Conversely, if the fetus receives impressions from and through the womb that it perceives as "hostile" and threatening to its survival, the potential for trust and self-love may be overshadowed by the potential for feelings of mistrust and lack of self-worth. It is most likely, and perhaps inevitable, that we all have had both types of experience to some degree. The development of love, trust, and fundamental attitudes toward oneself and life will of course be profoundly affected by experiences later in life, by the kinds of experiences that are dominant, and by our degree of sensitivity to these experiences. But it is in the womb that our potential to form loving and trustful relationships has its roots. Basic patterns of response begin to develop out of sensory information received in utero. This information is received and perceived as sensations of touch, movement, rhythm, and vibrations of light, sound, and thought, all of which are affected by the emotional energy of the mother and her environment.

To move freely and fully from one stage of development to the next higher one, the first stage needs to be experienced as a secure and supportive ground from which to take our next step. The developmental plan will unfold whether or not we have completely and securely experienced a certain phase. Whether we are able to take the next step fully or partially, with confidence or anxiety, or not at all, we will still be drawn unfailingly on toward higher levels by the promptings of the evolutionary "master plan." In going back to reexperience the movements, sensations, and "mind" of the early Navel Radiation pattern—to the extent that we as adults can do this—we have a chance to complete a stage that was incomplete before, or reintegrate where connections have been disturbed subsequently. Experiences of positive support, relationship, integration, and pleasure can be encouraged and strengthened.

By contemplating, embodying, and exploring the actual form and movement of this pattern, we begin to have some idea of the "mind" it evokes. Movements flow arrhythmically between flexion, compression, and integration of the limbs into the center, and extension, suspension in space, and opening out from the center. The limbs move together and apart, they touch and separate. They touch the walls of the womb and float suspended in fluid, discovering boundaries and the sense of boundaries dissolved. Movement is to and from the center, equally in all directions, in and out. In this "dance," movement embodies the interaction of the individual parts within a context of unity and wholeness.

The "mind" of this pattern has a quality similar to the cellular "mind" but involves the primitive beginnings of a sense of differentiation. Openness, spaciousness, receptivity, communication—these interchange and merge with boundaries, limitations of space, enfoldment, and self-containment. There is an experience of integration, wholeness, oneness, and unmanifest but infinite possibility. And this is the condition of the fetus, full of unlimited and not yet realized potentiality, as it approaches readiness for birth. The nearly newborn is uniquely and remarkably powerful, pregnant with itself, and at the same time totally vulnerable in its helplessness. It embodies an essential paradox of life.

The simultaneous experience of duality and nonduality, a fundamental condition of human existence, is already implicit at this stage. In the "mind" of this pattern we may play for a while on the threshold of consciousness, where all is possible but not yet manifest.

Chapter Three

Entering the World:
Prebirth and Birth

Every act of creation has a birth, a moment when the created comes into the light of day and can be seen, heard, felt, touched, and will also in some way see, touch, and feel the world it enters. Having come this far in the process of writing this book, I realize that it had its origins, its conception, at some time in the past, a seed of an idea invisible then to my conscious mind. As I learned, practiced, and shared with others the things I am now writing, I was not at first aware of them coming together in the context of a book; each was an individual, though related, event and experience in my consciousness at the time. As with the embryonic growth of a child, it is only at the time of birth that the new context emerges into view. However, during the long period of gestation the new form was being created within the dark depths of consciousness.

The birth comes as an act of necessity, a fulfillment of the process gone through up to this point. When the conditions are ripe the baby is pushed, and pushes itself, out into the world; the infant becomes the new context for the coming together of another whole level of learning experience. So too, on up the spiral of development, will each new stage necessitate such a birth, a separating from the old state and emergence into a higher state of functioning and consciousness. Each transition is an act of creation: a new form of life, and context for life, is born.

Birth itself is the first wholehearted act of will of this new being, a thrust towards greater consciousness and autonomy, a reaching toward a higher level of evolution. As with every act of will there is a choice. The child may recoil from or fight against the forces of life impelling it onward and may experience these forces as life-threatening. The contracting womb, for instance, might be felt as though it will crush and overwhelm the tiny delicate life within. Such intense fears are both understandable and real. Birth is a death of the old state of being, necessary for the transformation to take place, and is also a passage beset with dangers to the lives of both mother and child. Birth and death are always present together. The untried will of the child-to-be may also choose to cooperate with the greater will of nature and heaven, nature propelling it and heaven drawing it onward. Such patterns of resisting, rebelling against, or surrendering to and cooperating with the laws of growth and change we will surely recognize in the many transitions we are called to make throughout life's journey. Stanislav Grof, in his pioneering work with altered states of consciousness, has explored the relationship of prebirth and birth experiences to personality organization and patterns of transformation within the psyche.[1] His model of the Basic Perinatal Matrix closely parallels the stages of movement experience, mind states, and development observed and described in Body-Mind Centering work.

The great event of birth demands an immense act of will to initiate the powerful physical effort required on the part of the birthing child. Extraordinary strength is called forth in one so small to meet this struggle between life and death. This effort is vital to the child's emerging consciousness of itself and to its potential for sensory-motor and intellectual development. Learning, and the concomitant growth of intelligence, depends on the number of connections made between cells of the brain and nervous system as a whole, rather than on the number of cells themselves. During periods of normal stress or effort (which is at times both natural and necessary), such as we may experience in our attempts to master a new skill, great numbers of new connections are made between the cells of the nervous system. This increases the potential for sensory, physical, and mental learning. Joseph Chilton Pearce states:

> [B]irth stress prepares the brain and body for massive new learning. A general alertness, new brain connections, and new proteins are provided for the greatest movement from the known to unknown ever to be undertaken.[2]

Essential to the completion of this period of potentially dangerous and stressful activity is the subsequent experience of its counter-state of rest. A sense of relaxation and security follows and the birth process is fulfilled, physiologically and emotionally, when the child emerges and is returned to rest and be held by its mother. All too tragically in modern times, for us as individuals and as a society, this process is interfered with in a number of often unnecessary ways. And so the birth experience is incomplete, mother and child unfulfilled in the transformation they have labored for, and a state of unresolved stress sets up in the child. The balance of tension and relaxation, so necessary for health and well-being, has already been disrupted.

However joyful or traumatic has been the infant's entrance into the world, it is now alive, here on earth, and will be drawn inevitably through the next stages of its development. Its purpose now is to begin the unfolding of its potentiality, the embodying in life on earth of that which has evolved within the hidden confines of the womb, the original timeless and invisible universe. There is a basic ordering to these next stages of growth, a sequence of transitions that is largely predetermined and universal to humanity, its messages already encoded within the DNA. The nervous system will direct the expression of this ordered process of movement development through its communications with the body as a whole. The infant child will progress through these stages at approximately determined times in its early life, whether or not previous stages have been fully experienced and mastered.[3] Obviously the more completely fulfilled the foundations are, the more stable a base for future growth they will provide.

Within the basic plan there is of course much room for individual variations, and also adaptations or compensations that can usually be made to deal with inherent or functional weaknesses. Therefore, the child's process may appear to follow a very unique pattern. The uniqueness, however, is underlaid by the universality of the fundamental process of development. It is also interesting to observe that individual cultures adopt their own child-rearing practices and attitudes, and the differences in movement, perceptual, and social patterns present in the culture seem to reflect to some extent these cultural patterns in child care and development. (Whether the actual genetic structure may change with time, in line with these variations, and so the evolutionary plan itself adapt to the individuality of the culture, is an interesting line of inquiry, though beyond the scope of this book.) Let us now look further at the underlying process of development through the child's movement, as we observe it during birth and the first year of life.

The Mouthing Pattern

What we observe developing is the ability of the infant to initiate its own movement in a purposeful way, with conscious intent, and so move its own body through space or change its environment in response to its own actions. This can occur as the nerves that activate the muscles needed for a particular movement myelinate. These fatty myelin sheaths protect the nerve fibers and enable the stimulated muscles to coordinate movement patterns that are already latent within the nervous system, by greatly increasing the conductivity of the fibers.

As we saw earlier, the dominant movement pattern in utero is Navel Radiation, with movement organized around the navel center. In this movement there is little if any conscious intent in relation to the environment; the fetus moves according to implicit purpose and design. While this form and pattern of Navel Radiation dominates, it also provides a foundation for the fetus to go through several stages of transformation in preparation for birth and life beyond.

The nervous system develops into a spinal cord with the brain, its center of coordination and control, at one end. The mouth and sensory organs also develop in the head. The mouth, which is of primary importance immediately after birth, dominates this stage; several nerves that receive information from and activate the muscles of the mouth are the second group of cranial nerves to myelinate. The first to develop is the vestibular nerve, the perceiver of movement. This happens in utero, and the fetus can in fact find and suck its thumb, in practice for the nursing, breathing, and vocalizing activities which will be needed immediately after birth.

The Mouthing pattern is seen in the tunicates, such as the sea squirt. These water-dwelling, sac-like creatures have a large mouth at the top through which food and sometimes also wastes pass, as they sway from their mooring on the ocean floor. Some also give birth through the mouth. In the tunicates, the center of control has in effect moved from the actual center, as in the starfish, to one end of the organism (Fig. 3.1).

The adult tunicates have a sedentary nature. If they do move from one location to another, they tend to drift with the currents rather than propel themselves in a directed way. Their young tadpole-shaped offspring, however, are able to swim freely, guided by the head and propelled by the long tail through the water. Here we see the transition from "being" to "doing" nature: the stationary mother gives birth to the activity of her offspring. It is this same movement of thrusting and reaching forward through the head that will take the infant from its life of being in the womb into its life of doing in the world. (Fig. 3.2)

Figure 3.1 The tunicate or sea squirt embodies the Mouthing pattern in its form and function.

46

Figure 3.2
The tunicate offspring has
a tadpole-like appearance.

One theory of evolution proposes that some species are able to bypass their adult stage and reproduce while still in the larval stage, a process called pedomorphosis. Their own offspring will take on the characteristics of this early stage of development when mature, and a new form of life will have evolved. The tunicates are thought to be one such species, and so exhibit an important transitional stage in evolution.[4] The young display in simplified and prevertebrate form the basic characteristics of more complex spinal creatures. The mouth of the "mother" species is still prominently located at the front of the head of the young and guides both its head and body forward into movement as it reaches into the environment in search of food.

For the infant the mouth is equally significant before, during, and after birth, both in the search for food—its mother's nipple—and as a base for further movement and vocal development. Motivated by the need to find food to survive, the mouth will reach towards the source, drawing the head with it. (Fig. 3.3) Movement initiated at the mouth in this way will reverberate down the spine. The action is initiated at the jaw (the temporomandibular joint); while the chin or lower jaw is partially resting against the mother's breast for support, the upper jaw is primarily responsible for the action of reaching for and grasping the nipple. In this way the whole head rocks forward and back with the rhythm

Figure 3.3   In feeding, the baby reaches for and grasps the nipple with its mouth.

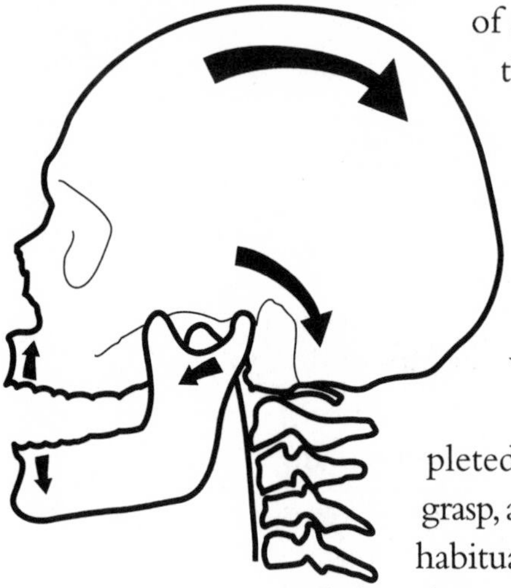

Figure 3.4
The action of the temporo-mandibular joint (TMJ) initiates a rocking movement of the head: this impulse levers into the vertebrae of the spine to initiate the Spinal Push from the head.

of sucking; as the mouth opens, the movement of the skull levers into the first vertebra and then sequentially down the whole length of the spine to its tail. As we will see later, this action underlies the first Push pattern from the head, down through the spine. (Fig. 3.4) As the mouth closes and the skull circles up and forwards again, the whole spine is released and lengthens in response.

If the closing phase of this action is not completed, the mouth will be left slightly open, unable to grasp, and the head dropped back. If this pattern becomes habitual, tension in the muscles of the neck will develop and the release of the spine upward, which underlies the action of reaching with the head, cannot occur. This very common pattern underlies many neck, shoulder, and back problems, and may originate even before an infant begins to nurse, right from the moment of the first breath. Many adults of current generations suffered as newborn infants the brutal practice of being suspended upside down and slapped in order to draw the first breath. Such trauma will throw the infant into the Moro (startle) reflex: the mouth opens to release a cry, the head and arms are thrown back, and the muscles of the back contract. This reflex creates width across the front of the chest as the arms open and the spine arches back, so that the baby can then reach out more fully to embrace a source of comforting contact. In this first traumatic experience of life, however, the baby, hanging by its feet in the air, has no such source of safe contact to reach for and embrace. The position in which it is held does not allow the infant to complete the closing phase of the reflexive action. Every subsequent shock would tend to reinforce the incomplete pattern, for the closure and embrace needs to be experienced fully before the reflex can be integrated. When we think about the epidemic of back problems in our culture we would do well to consider this pattern among the possible contributing causes.

For the infant, sucking is a whole-body activity which creates a base for movement in space through the spine. If you watch a baby feeding, you will see a great deal of activity in the spine and organs of the abdomen and pelvis; the hands and feet will also be actively involved. As the baby sucks breast or thumb, there is stimulation to its vestibular nerves (see Chapter Two), the bones of the skull, and the glands of the head. Sucking is also an active searching, reaching, and grasping activity that lays foundations for the future patterns of reaching and grasping through the hands, the feet, and the other senses. The psychological implications of this pattern concern the individual's ability and willingness, at a core level, to reach out for and take in what she or he wants and needs from the environment. This is the active side; the receptive aspect of this pattern concerns issues of satisfaction or frustration of these needs by the external environment.

An infant who is bottle-fed may not have quite the same experience of active searching, of learning to fend for itself and make this step toward independence and the sense of real personal power. The bottle-fed child may continue to experience the world as coming toward it, fulfilling its needs, with less sense of being actively involved and in control of this life process. This more passive experience could contribute to feelings of contentment or complacency and a mistaken sense of power, on the one hand; or, conversely it may engender a sense of being helpless, not in control, or at the mercy of the will of the environment. This is not to say that both frustration and satisfaction cannot be experienced at times in both methods of feeding. A bottle-fed child, however, is more likely to miss out on an important movement experience and may therefore not embody as fully as the breast-fed child the "mind" state and energetic experience associated with this pattern.

If the Mouthing pattern does not fully emerge or is not practiced completely enough, the integration of the head into the spine in sequential spinal movements is inhibited. The rocking and turning movements of the head will initiate primarily in the muscles of the neck, instead of those of the mouth and face. This causes tension to build up in the neck and results in a disconnection of the head from the rest of the body in movement. Early patterns of locomotion are integrated whole body movements, initiated at the extremities. The first patterns to emerge happen through the length of the spine, initiated first at the temporomandibular joint in the movements of the mouth. These movements ripple down the spine to the tail; then the tail initiates the movement which sequences back through to the head. We see this reflected in the movement patterns of the fish, snake, and inchworm or caterpillar. Because the nervous system is organized with its center of control in the head, each pattern will be initiated first at the upper end of the body, then the lower. Muscle control and coordination of the limbs also occurs first at the center of the body around the spine, and progresses outwards toward the extremities. Development, therefore, also proceeds from center to periphery.

The Pre-Spinal Pattern

Along with the Mouthing pattern, the Pre-Spinal pattern also develops in utero and is a precursor and a support for the actual Spinal patterns. The notochord, a flexible "stiffening rod" that runs through the length of the torso, appears at a certain stage of fetal development. It lies between the spinal cord at the back and the digestive tract at the front, and serves to differentiate these two. During a later stage of fetal development the notochord will be absorbed and the bony vertebral column and skull will eventually take its place. Remnants of the notochord remain in the center of the discs between the vertebral bodies.

The Pre-Spinal pattern involves moving from both the digestive tract (mouth to anus) and the spinal cord (brain to conus medullaris); this pattern relates to the level of the notochord or "soft spine" that underlies the "hard spine" of the vertebral column. The Pre-Spinal movement, which has a flexible, serpentine quality quite distinct from the quality of moving from the spinal vertebrae, establishes a base for sequential movement through the spine and underlies the integration of the head and spinal column. The movement of the "soft spine" can be clearly seen in a small infant before the Spinal patterns have been fully integrated; it may be initiated through both the spinal cord and brain, and through the digestive tract. (Fig. 3.5)

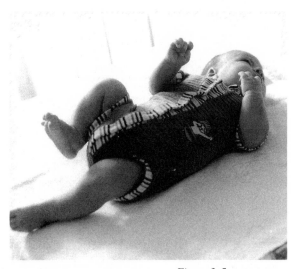

Figure 3.5
The Pre-Spinal pattern—the "soft spine" of the spinal cord or the digestive tract initiates the movement.

This stage of development relates to the lancelet amphioxus, which is more or less fish-shaped but does not have a bony spinal column. Like the young tunicate, its mouth is prominent and leads it forward through the water. The head is not separate from the rest of the body and does not have special senses other than a mouth. The body is segmented and movement is sequential, guided by the head and propelled by the tail. The adult lancelet amphioxus has a notochord with a nerve cord lying above it and a digestive tube below. In terms of evolution, this species could be looked upon as a transition stage between the invertebrates and the vertebrates, having many of the characteristics of form of the vertebrates without having an actual vertebral column. Like the earlier invertebrates, it also lives in water. (Fig. 3.6)

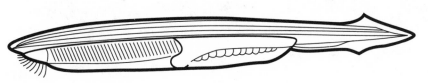

Figure 3.6
The lancelet amphioxus displays the Pre-Spinal pattern in its structure and movement.

51

The Spinal Patterns

Let us now return to the birth itself. The birthing movement is primarily a pushing action initiated by the head; the Mouthing pattern underlies this action of the head, which reverberates in a sequential movement down to the tail of the spine. This impulse as it meets the contracting walls of the womb will create a responding push from the tail (and feet) back along the spine to the head. We call this movement pattern a Spinal Push, initiated first by the head, then the tail of the spine. The spinal patterns have been present in utero and are the dominant mode of locomotion both during and immediately after birth. The Spinal Push[5] reflects, phylogenetically,[6] the movement of the inchworm[7] (Fig. 3.7) and can be recognized in the young infant's ability to wriggle itself to the far end of its crib before any controlled motor ability has developed in its arms and legs to assist this movement. The infant "inchworms" by alternately flexing and extending its whole spine, pushing the body gradually along; the limbs are reflexively involved in this movement. (Fig. 3.8) The Spinal Push, of the head in particular, also provides important stimulation for the pineal and other glands of the head.

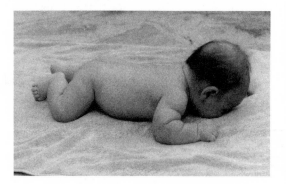

Figure 3.7
The movement of the looper or inch-worm displays the qualities of the Spinal Push pattern.

Figure 3.8
The baby first begins to propel himself along using the Spinal Push pattern; the Mouthing pattern supports.

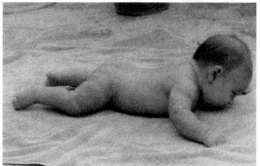

As the head of the birthing child pushes into and through the birth canal and the tail of the spine and the feet respond by pushing against the contracting walls of the womb, the push of the head transforms into a reaching through to the new world outside. As the head breaks through and pressure is released, the head literally reaches forward and pulls the body with it. The actual entry into the world is the infant's first active reaching toward a higher level of existence. With this movement, the infant's first response of "yes" to the will of life and heaven, drawing it onward, is completed. The Spinal Reach and Pull pattern, initiated here at the head, is best reflected in the movement of the fish, in which the flicking of its tail creates a force to push forward as the head reaches beyond and pulls the body through the water. (Fig. 3.9) We will see this pattern recurring at each transition from one level to another higher one throughout the infant's development, for it is the reach of the spine from the head or tail which initiates such transitions. (Fig. 3.10).

Figure 3.9
The Spinal Reach and Pull pattern is seen in the movement of the fish.

The birthing child has pushed and pulled itself through the birth canal from the watery world that has been its home into a world of earth and air and "other." It has now truly separated itself and become its own person, though as yet it is not fully conscious of this transition. During birth the Spinal patterns crystallize and are integrated as a functional mode of activity. They are the primary means of locomotion in the newborn and young infant, and they also form the foundation or matrix for the next stage of development. These patterns will be present throughout life, underlying all movement and perceptual activity.

Figure 3.10
The Spinal Reach and Pull pattern brings the baby to a higher level; the senses are actively supporting.

The Spinal Push patterns develop the "mind" of inner attention. The slight compressive force that passes through the spine

53

in the Spinal Push patterns establishes the vertical axis of the body through proprioceptive stimulation; an integrated vertical axis gives an energetic basis of support for the experience of self, of "I-ness," which is beginning to develop. The vertical posture and self-awareness are related characteristics of humankind. At this early stage the infant's movements in space are primarily in the horizontal plane, along the floor or other supporting surfaces. It is essential that the sense of the vertical axis in the body develops first in this relation to the earth, in order that a clear and supported relationship with gravity be established through the spine before reaching to the upright posture. The Spinal Reach and Pull patterns give the sense of elongation of the vertical axis into space, develop the "mind" of outer attention, and underlie all changes of level. The head reaches and pulls forward and upward, perceiving the space beyond itself in front, while the tail of the spine reaches and pulls the body back, down, or up, perceiving the space behind and beyond.

Led by the head, movement of the spine in all directions can be initiated by the infant through stimulation of the special sense organs (mouth, nose, ears, and eyes), and gentle touch to the skin of the face. This creates a base for the development of body movement in the three basic planes—vertical, sagittal, and horizontal—and the diagonals which combine all three dimensions.[8] Practicing these movements therefore creates the foundation for

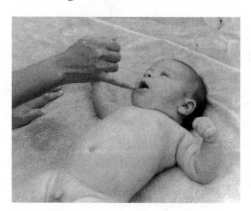

Figure 3.11
Stimulation of the senses initiates turning of the head—the Rooting Reflex.

movement in space through all dimensions and planes. The infant will first respond to touch and stimulation of the mouth and sense of smell; this is called the Rooting Reflex, and it underlies later responses through the higher senses. (Fig. 3.11)

Sound and visual stimulation will provide increasingly more significant motivation for movement over the first few months of
life as the functions of these organs develop and coordinate with
bodily activity.

Exploration: Mouthing and Spinal Patterns

Practicing the exercises described below will help create a clear
foundation for the more advanced movement patterns to follow.
The early patterns are coordinated by lower brain areas (see Table
2, pages 96–97); in turn, the respective areas of the brain are stimulated by the performing of the early movement patterns. This
helps to regulate the related functions of the lower brain, and frees
higher brain areas to coordinate more sophisticated movement
patterns. Coordination of the early patterns by the lower brain
areas should take place automatically, without conscious control.
If this is not happening freely, for example because of damage to
or dysfunction of areas of the lower brain, higher brain areas will
be required to coordinate basic movement responses. This means
that less attention and energy is available for the perceptual, intellectual, and creative processes of which the higher brain is capable.

We can return with conscious awareness to clarify the early
movement patterns and in this way free more of our energy and
conscious attention for creative, social, and intellectual activities.
Stimulating the special sense organs of the head facilitates a clear
and effortless movement and awakens the senses to a more active,
expressive, and direct contact with the environment. The freeing
of the temporomandibular joint in the Mouthing pattern is of
particular significance as a base for reaching and grasping activities initiated by the other senses and the hands and feet; this freeing can release a tremendous amount of energy held in the spine
and soft tissues of the body.

Exercise 1

This first exercise is best done lying down on your back, as you will be more relaxed in this position. The movements may then be tried in different positions, such as lying on your front, sitting, on hands and knees, or standing. It is helpful to have a partner giving you the stimulation and guidance for the movement, but this exercise can also be done alone.

When you feel relaxed, gently stroke the area of the face around the mouth, lightly brushing outward along the cheek, and allow the head to turn, following the direction of the touch. The touch stimulates the muscles around the mouth to initiate the turning movement, as in the Rooting reflex of the infant.

The back of the head should also make a turn, equal to the movement of the face, so that the head rotates clearly, in the horizontal plane, around its vertical axis. Feel how this movement sequences through the neck and could initiate a rotation down the length of the spine. (Fig. 3.12)

Repeat this movement, initiating by stroking first the area just under the nose, brushing outwards; then across the cheeks toward the ears, and finally from the skin around the eyes out along the

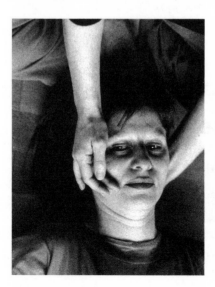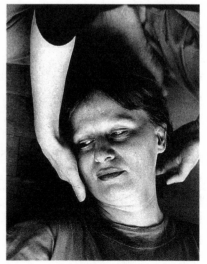

Figure 3.12
Initiating head turning
through the senses.

upper cheekbones and the temples. This gentle touch awakens the sensitivity and responsiveness of the nerves of the skin, which stimulates movement in the muscles of that area. Also let the senses of taste, smell, hearing, and sight lead you to turn toward scents, objects, and sounds used for stimulation. In this way the process of sensory-motor integration, which we first saw emerging in utero, can be facilitated. (Fig. 3.13)

Once the turning movement feels easy and smooth, let your open eyes actively follow an object or your own hand passing from side to side in front of your face. Allow the eyes themselves to initiate the rotation, followed by the head, and allow the movement to sequence down through the spine. In coordination with the hand reaching in the same direction as the rotation of the head, this initiation can take you into movement through space or into a sequential rolling through the whole spine. This process underlies the Reach and Pull patterns. (Fig. 3.14)

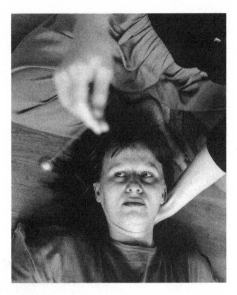

Figure 3.13
*Visual stimulation—
tracking underlies hand-
eye coordination.*

Exercise 2

First find a comfortable and relaxed position, perhaps lying on your side, then place your thumb in your mouth, touching the roof of the mouth where the hard and soft palates meet. This stimulates the sucking reflex. Allow the head to rock slightly forwards and back as you suck, so that the movement is not solely in the lower jaw. This action should happen naturally if you relax and try not to inhibit or interfere with the movement. The movement occurs at the temporomandibular joint, just in front of the ears, and is the initiator of movement

Figure 3.14
*Reaching through the
eyes and hand initiates
rotation through the
whole spine.*

through the spine. The patterning of movement at this joint therefore has implications for the alignment and movement patterning of the whole spine.

Feel how this movement creates a slight flexion and extension in the neck, beginning where the skull rocks on the first cervical vertebra. You may be able to feel an impulse from this movement traveling down along the spine, and also through the digestive tract in front to the abdominal and pelvic organs.

Exercise 3

Then take up a position on the floor, face down, supported on the forearms and forelegs; the knees and elbows are flexed and wide apart, and the toes and fingers touch each other, forming a diamond shape on the floor. Sit back into your heels and rest your forehead on the ground. Rest here for a moment, allowing the muscles of the back and the organs to relax.

Initiate a push forward from your tail (Fig. 3.15). (Your feet and forelegs may be used to assist a little with the initiation of the push, but keep the focus and primary impulse on the tail of the spine.) Let the impulse sequence through the vertebrae of the spine, each pushing gently into the one above, right up to your head. The spine will travel forward and curve into a deeper flexion as the head rolls on the floor to let the weight pass through the top of the head. The pelvis is now high off the ground. This is the Spinal Push initiated by the tail.

Figure 3.15
Initiation of the Spinal
Push from the tail.

Slightly open the mouth at the temporomandibular joint, as in the Mouthing pattern above, rocking the head back a little. You will feel the top of the head press slightly into the floor as you do this, and the skull will also begin to roll back and lever into the first cervical vertebra. Allow the movement impulse from this gentle push and rolling of the head against the floor to sequence back down through the vertebrae of the spine to the tail. (Fig. 3.16) This takes your spine backward in space, until you are again sitting on your heels with your forehead resting on the floor and the spine lengthened, as at the beginning of the exercise. This is the Spinal Push initiated by the head.

As you repeat the push from the tail, feel the closing of the mouth at the temporomandibular joint underlying the rocking of the head forwards. The neck and spine flex into a curve and the weight travels forwards onto the top of the head.

Alternate these two movements a few times, slowly at first, feeling the underlying action of the Mouthing pattern.

Then sense into the center of the back half of the spine, and breathe there into the spinal cord and also into the brain; repeat the movement initiating from here. Then breathe into the organs, sensing the length of the digestive tract, and move through the organs as you do the exercise again. This is the Pre-Spinal initiation. Notice any differences in the feeling and quality of the movement.

Figure 3.16
Initiation of the Spinal
Push from the head.

Exercise 4

When you can feel the integration of the whole spine and head in these movements, you can try the Spinal Reach and Pull patterns. Begin in the same position as for the Spinal Push patterns, but this time allow an active reaching through the top of the head to initiate the movement, pulling the head and spine in a long curve forwards along the floor. The chest, abdomen, and thighs will at first be drawn out close to the floor, but as the head reaches up and forward the whole spine will be pulled into extension and onto the support of the hands. The support and initiation of the eyes are important in this movement. (Fig. 3.17.)

Still keeping the eyes attentive and outwardly focused, reach back and upwards through the tail (Fig. 3.18), pulling the spine up onto the support of the hands and knees into the quadruped

Figure 3.17
Spinal Reach and Pull
from the head.

60

Figure 3.17
(continued)
Spinal Reach and Pull
from the head.

Still keeping the eyes attentive and outwardly focused, reach back and upwards through the tail (Fig. 3.18), pulling the spine up onto the support of the hands and knees into the quadruped posture; you can then pull back further into the starting position and repeat the Reach and Pull through the head.

Alternate these two movements several times, until you find an easy flow and a sense of lightness and aliveness throughout. As

Figure 3.18
Spinal Reach and Pull
from the tail.

with the Spinal Push, you can also explore initiating the Reach and Pull movements in the spinal cord and digestive tract.

Chapter Four

The Spiral of Growth: Moving on Land

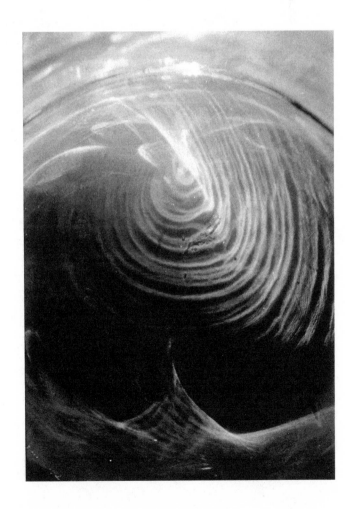

The human body is made up of about 70 percent fluid; each living cell consists of about 80 percent water. It is one of the five basic elements essential to all forms of life and out of which all life is created.[1] In his book *Sensitive Chaos,* Theodor Schwenk writes:

> Wherever water occurs it tends to take on a spherical form. It envelops the whole sphere of the earth, enclosing every object in a thin film. Falling as a drop, water oscillates about the form of a sphere. A sphere is a totality, a whole, and water will always attempt to form an organic whole by joining what is divided and uniting it in circulation.[2]

The cell itself is essentially spherical in form, following this basic tendency of water, and hence all living matter, to unite its parts into a whole. We can call this a law of nature.

A second law of nature that acts upon all matter, organic and inorganic, is gravity, which irresistibly pulls everything with substance towards the center of the earth, bonding all that is of the earth to her. The interaction of these two tendencies creates movement in the form of a downward spiraling. Everywhere in nature this spiral form is present: the shell of a snail, the currents of a meandering stream or a flow of air, the bark of a tree, the unfolding of leaves around the stem of a plant, or the opening of rose petals. As the cells of the human fetus multiply and arrange themselves into distinct structures within the body, they also follow this patterning: the striations within bone tissue spiral downward, enabling the bones to dynamically transfer the force of weight falling to the earth; muscles wrap around the bones in a continuous network of spiraling movement; the heart itself is a muscle which spirals in and around itself; the fluids of the body, too, flow in currents, waves, and gushes, reflecting all the myriad spiraling forms of nature's rivers, oceans, and waterfalls.

Inorganic matter is bound to obey the law of gravity; a stone cannot, of its own accord, rise upward, defying gravity, nor can a stream flow anywhere but toward the earth's center. But in the organic world a counterforce occurs, a movement upward, which is the force of antigravity, or levity. Wherever there is matter that has life, feeling, and consciousness, however primitive, there is the possibility of growth and movement upwards as well as down. This movement is also spirallic, as the tendency to grow toward the wholeness of the sphere is still present. While the earth in which a tree grows will always settle downward with its own weight, the tree itself, having its own life and feeling and responsiveness to the cycles and changes of its environment, will spiral its path in both directions, reaching up toward the light and warmth of the sun as well as down to the source of nourishment in the earth.

And so it is with us. As human beings of both matter and spirit or consciousness, we can live and grow and move on the earth through a dynamic interplay of these two forces of gravity and levity, both bonded to the earth and growing upward to the heavens, connecting the two within ourselves through our unique relationship to each. The very structures of our body come into form through these spiral patterns of fluid movement. Our ability to move upon the earth is mastered in infancy through a spirallic process of development. And the natural evolution of an individual's consciousness throughout life's journey of learning can also be compared to a movement that spirals both up and down as new dimensions of one's being are embraced.

The Developmental Process Underlying Movement

As mentioned earlier, movement develops through a series of pre-determined stages. A movement belongs to a specific pattern based on the way in which it is initiated and how it sequences through the body. Early movements initiate at the extremities and sequence through the whole length of the body to the opposite extremities, creating an integrated whole body movement. The initial impulse comes in response to the external environment and the inner needs of the infant; this impulse creates a ripple of action throughout the body, causing it to move away from or toward the stimulus.

This particular way of identifying developmental patterns is, to my knowledge, unique to the Body-Mind Centering approach. The neuromuscular coordinations—spinal, homologous, homolateral, and contralateral—are recognized by traditional authorities on movement development, but the related initiation and sequencing of movement through the body are not.[3] The primary stimulus to intentional and purposeful movement comes from the infant's bodily survival needs for food, breath, contact, and comfort. As we have seen, the nerves that activate the muscles of and around the mouth have already myelinated by birth, and in the normal course of development the newborn infant will respond to a gentle stroking of the skin around the mouth by turning toward and grasping with its mouth the object of stimulation, or turning away from the object if it is not desired at that moment. This grasping with the mouth activates the sucking reflex. The action of turning and reaching with the mouth is followed by activity in the muscles of the neck; the contractions will eventually travel sequentially throughout the length of the spine in a coordinated spirallic rotation, head to tail, as myelination of the spinal nerves is completed. Similarly, opening and closing of the mouth will initiate extension and flexion of the neck and spine. As described earlier, the Mouthing pattern underlies the Spinal Push and Reach/Pull patterns initiated by the head. The sense of smell is also highly developed at this early stage, and the infant will respond in a similar way to the smell of its mother's body and her milk, initiating movement from the area of its nose. Once myelination of the nerves that activate all the spinal muscles has occurred, the Push and Reach/Pull patterns can also be actively initiated at the tail of the spine.

If movement does not initiate at the extremities, it will not sequence through the whole body but will remain a more or less localized activity. This can set up a pattern of tension where energy is locked into the tissues of that area. If the muscles of the mouth, face, and the other senses do not initiate the movements of the head, for example, such movements will be a localized activity of the muscles of the neck, causing tensions in the neck and a disconnection between head and torso. When initiated and sequenced clearly, the developmental patterns embody in movement the internal connections between body parts that we explored in the Navel Radiation pattern. These connections underlie the ability to articulate the separate limbs independently while maintaining the inherent unity and integrity of the whole.

The way in which we initiate movement also makes a statement about our relationship to our environment. Life is a continual interaction between our inner and outer worlds, and a healthy relationship between the two requires appropriate responsiveness to both. As infants, we first learn to distinguish inner from outer, "I" from "other," through the developing experience of our body boundaries.[4] The skin is the primary boundary, and physical contact is essential to the infant's developing sense of an "I," the quality and frequency of such contact being determining factors in the emergence of a healthy, stable, and integrated sense of self. Out of this sense of "I-ness" the child will gradually learn to identify, and willfully act in response to, its inner needs and wishes and the demands of the environment.

It is through the body extremities and the special senses of the head that we meet the environment most directly; through their actions we can also interact with and respond most clearly and effectively to it. So when we initiate our movement peripherally there is clarity and directness in the form our movement takes, its expression in space, and in the quality of communication between ourselves and the world outside. If movement is habitually initiated more centrally and is localized in a limited body area, our focus tends to be inner-directed and we have less attention available for external interaction.'

We are not looking to develop one mode at the expense of the other, but rather to find a balance between inner and outer, between movement initiated centrally and peripherally. Both are natural and appropriate in particular circumstances, and the balance between the two will be different for each individual personality and temperament. However, I emphasize here the initiation of movement at the body extremities because it is this that many of us may have never fully developed, or may have lost, as growing children and adults. Fear or the continual frustration of needs and desires in the infant and young child may cause a withdrawal from direct interaction with the world and may turn the focus of attention too far inward. We then cease to push away or reach out and move toward with all our heart and strength. The ensuing loss of vitality and presence can take on many forms of emotional and physical disturbance, sometimes much later on in life.

The infant's first needs are for food, breath, warmth, comfort, and physical contact. When such essential needs are adequately met and a safe enough holding environment is created, the infant is enabled to begin its movement toward independence and creative interaction with the world. First it must learn about the earth onto which it has been born, and how to move upon this earth, so that it can be instrumental in getting its own needs met. During the period in the womb, a foundation for this bodily understanding of the earth's force of gravity has been developing, and now the infant will learn to master its responses to this force in a purposeful way.

After birth the infant, provided it feels secure enough, can surrender or yield its weight to the force of gravity, bonding physically to the earth and to its mother through whole-body contact with her supporting surfaces. It cannot yet lift itself up, but this yielding and releasing of its weight down to meet the earth is the first step in discovering both the force of antigravity and its own mobility. As the weight presses into the ground and is received, force meets force; the ground yields and responds with an upward thrust which travels through the infant's body. Gradually, as nerves myelinate and the muscles of the arms and legs begin to coordinate and strengthen through practicing movement, the infant is able to push its body up off the ground and travel through space. In this way the push patterns are initiated, first from the head and tail as described in the last chapter, then from the forearms and hands, and finally from the forelegs and feet as these make firm contact with the ground.

The Push Patterns

At each turn of the spiral, each transition from one stage to the next, an act of pushing against the old matrix of support is required to initiate the process of change. The action of pushing has an internally directed focus, but this is not a withdrawing from the world; it is an expression of great power, strength, and presence. As the child pushes against the ground the energy of the earth flows up into its body, strengthening and nourishing it. The body tissues are subtly compressed by this mutual action of body weight meeting a resistant surface, and this enables the child to feel and become aware of its physical substance, weight, and presence. In the pushing patterns the child's attention is primarily involved with inner sensations of weight, gravity, balance, and movement, and with its evolving sense of self and boundaries. The pushing action levers the body through space with a movement that has a stable, weighted, strong, and self-absorbed quality.

The pressure that stimulates the push may be felt both physically and psychologically, internally and externally, in ways both obvious and mysterious. The meeting of the two forces and their response to each other is the necessary preparation for transition and transformation. Saying "No" to the old way precedes the welcoming of a new level. But even prior to this statement of "No" and the move towards greater independence is the universal "Yes" to life. The yielding of the infant to gravity to bond with the earth and mother both strengthens and is an expression of this "Yes." The bonding process and related state of preconscious unity, experienced in the womb, underlies and makes possible all future steps in growing and learning.

Rooted in this experience of support and containment, the infant's push emerges as an act of differentiation. Through it the child begins the transition from a state of being merged with another to one of greater independence and autonomy as an individual: I can push the world away from me, push myself away from the world, push in order to get a response from the world, push just to feel myself, or push through from one world to another. If that against which I push is both stable and responsive, I begin to establish my individual sense of self and come into relationship with the ground, force, mother, and social structures as I push. I can test out my strength and power and discover my uniqueness. The push may be a "No" to or a testing of the present order of things; through it I claim my individuality and independence. This established, I can again say "Yes" and bond at the next level to life and all that represents life to me at this moment.

There are risks in this willful act, for I may push myself away or push the other too far away, and so experience myself in isolation. Both sides must be willing to dance this dance. Pushing must be supported and balanced by yielding mutually. It takes courage from both to commit fully to this action. Life is committed to calling forth its child at birth, despite the danger that life may be lost in the process. Both mother and child are compelled to take up this challenge.

The Reach and Pull Patterns

Gradually, over the first few months of life, impressions taken in through the ears and eyes will begin to take on more importance in the overall development as the brain develops its ability to process this information into meaningful perceptions. As the infant becomes more secure in its environment and sense of self and in its ability to act upon the world through its own physical actions, it will develop a playful curiosity about this world. When its basic survival needs are being adequately met, attention will begin to turn more and more outward to an exploration of the world around it. This shift of attention from inner to outer environment and the increasing curiosity and desire to interact with the world will initiate the actions of reaching out, at first to the limits of the infant's own personal space.[5] We have already seen this action occurring in the reaching of the mouth toward the source of food; this establishes a baseline for subsequent active reaching and grasping actions of the hands and feet as well as through the other senses of touch, smell, hearing, and seeing. The individual's pattern of initiation at the mouth will be reflected in the way in which he or she reaches and grasps through the other senses and limbs.

When a firm foundation has been established through the development of the push patterns, and desire and curiosity in the child unite with the courage to extend out beyond its known boundaries, the hand, foot, head, or tail will reach through space and pull the whole body with it in a sequential flow of movement from the initiating extremity through the entire length of the body. During this action, a sense of space is created within the body tissues, imparting to the movement a quality of lightness and finely articulated control. The environment is the stimulus and support for such movement. When attention, focused on the object of stimulation, is aligned with the intention to move toward or away from that object, then movement happens with a clarity of coordination, directness, and the full energy of commitment and decisiveness. This we can see in the boundless vitality of a child at play.

The "mind" of the Reach and Pull patterns is one of outer focus and it expresses a quality of lightness, alertness, ease, and spirited spontaneity. However, if a sure sense of self and boundaries have not been established, the action of reaching out and moving through space can create feelings of disorientation and ungroundedness, a sense of being lost in space with no home ground to which to return. Many autistic children seem to display this tendency to an extreme degree. On the other hand, without the quality of the Reach and Pull patterns, movement may remain heavily earthbound and too self-reflective, lacking vitality and inspiration. In comparison to those with autistic behavior, children with Down's syndrome frequently display such a weighted quality in their movement and tend to be stronger in following another's direction than in initiating their own creative movement. When both patterns are familiar and accessible to us, we can move and act with greater choice and appropriateness. Ideally we would express an integration of both in our everyday movement.

The Sequence of Developmental Patterns

Let us now look further at the series of movement coordinations which, as we have already seen, are present within the nervous system throughout intrauterine development. Here the "shadow" of their expression has been developed and experienced in the Navel Radiation pattern.

Several principles underlie the process:

1. Push patterns precede and provide a grounding for the Reach and Pull patterns and can be returned to when there is difficulty in performing the Reach and Pull patterns.
2. A movement pattern is initiated first by the upper body extremities: the head or hands; then the lower: the tail or feet.
3. Balance in stance precedes balance in movement.
4. Support comes first from the ground, then from space.
5. Development of the limbs is from proximal (those parts closest to the body center) to distal (those parts farthest from the body center, the extremities).

Movement follows the mind intent. The presence of appropriate stimulation activates the desire, intent, and will to move, and then to master new levels of movement ability, embodying the potential latent within the nervous system.

The first movements that travel through space, as described earlier, are spinal—the wriggling and rocking actions through the length of the spine, initiated by pushing, develop into the reach and pull of head and tail as attention focuses more outward. As the head and tail are reaching out and upward, the arms and legs are developing their ability to support the weight of the body off the ground. This is a preparation for and transition to the next movement stage, the Homologous Push patterns. Here we see the overlapping of the development of the push phase of the Homologous with the reach and pull phase of the previous Spinal pattern. This overlapping occurs throughout the developmental sequence; the Spinal Reach and Pull patterns initiate every change to a higher level, and they support the development of the push patterns through the limbs.

The Homologous Patterns

The Homologous patterns are initiated by both arms or both legs together, and are reflected in the movement of the frog, rabbit, or kangaroo, for example. (Fig. 4.1) Playing leap-frog or diving into water, we are also expressing these patterns. They differentiate the upper and lower halves of the body and occur in the sagittal plane.

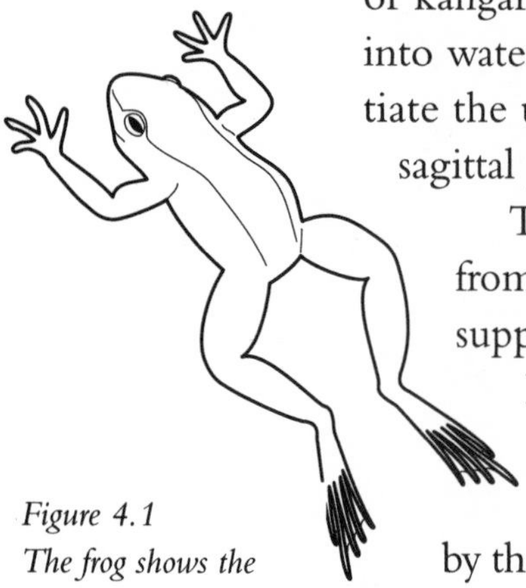

Figure 4.1
The frog shows the
Homologous pattern.

The first pattern to emerge is the Homologous Push from both upper limbs. As the head is lifted high onto the support of the elbows and forearms or right up onto the hands, and the tail extends back, a pushing action through the hands and forearms propels the body backward along the floor. (Fig. 4.2) This pattern is followed by the Homologous Push from the lower limbs: the two feet, forelegs, and knees push against the floor, taking the legs from flexion to extension. This propels the body forward and lifts the head up higher, from the elbows onto the support of the hands. Or the push from the lower limbs may continue through into a scooping movement of the hands against the floor, similar to the action of swimming the breaststroke. (Fig. 4.3) These patterns develop strength, gross body coordination, and extension of the limbs and spine; they express the mind of "inner intention." The achievement of balanced support on the forearms marks

Figure 4.2   The
Homologous Push from
the upper extremities.

the moment when, with the head lifted to vertical for the first time, the child can look directly ahead at the world and begin to recognize its individuality, its "I-ness" and humanness. This is the posture of the Sphinx, expressing the power, beauty, and mystery of nature and spirit embodied.

Figure 4.3 The Homologous Push from the lower extremities.

Once the limbs are fully extended and active to the fingertips and toes, the Homologous Reach and Pull patterns will begin to develop. An active reaching of the fingers of both hands together, along with spinal initiation supported by the senses of the head, will pull the body forward through space. This is the first pattern (after the birthing movement, or reach through the spine) in which we are able to extend ourselves, with full commitment, to reach beyond our personal space or kinesphere, displacing our own center of gravity. (Fig. 4.4) Once the initiation through the hands has integrated, the toes of both feet will also be initiating a pull backward; visual focus is still a support and initiation for the reach and pull backward. These patterns can be

Figure 4.4
The Homologous Reach
and Pull from the upper
extremities.

Figure 4.5
The young child learns
to master the Homolo-
gous Reach and Pull
pattern as she moves
from flexion to
extension of both arms
simultaneously; when
ready, this movement
will draw her out into
space.

seen in the infant's first sliding itself from a height to the floor, in reaching out from a sitting position onto all fours, or in spontaneously reaching out and throwing itself into your arms, trusting that you will catch it. (Fig. 4.5) The Homologous Reach and Pull patterns express the mind of "outer intention," and courage, commitment, and trust. This is where we first learn to leap wholeheartedly into the unknown territory beyond the safety of our personal space. It is important that the security of the holding environment does not abandon the child at this stage of growing independence and self-assertion; the "earth" must remain steady as the child leaps away and returns.

The Homolateral Push Patterns

Again overlapping with the development of the Homologous Reach and Pull patterns, the next Push patterns are also emerging. These are the Homolateral Push patterns, initiated first by the hands, followed by the feet. In these patterns the two sides of the body are differentiated. With the upper body now securely

supported on the fore-
arms or hands, the infant
begins to shift its weight
from side to side, and so
can initiate the pushing
action from one hand or
forearm at a time. The
impulse travels through
the spine to the leg on
the same side of the body;
the leg is then extended

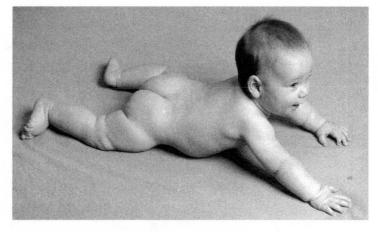

*Figure 4.6   The
Homolateral Push from
the upper extremities.*

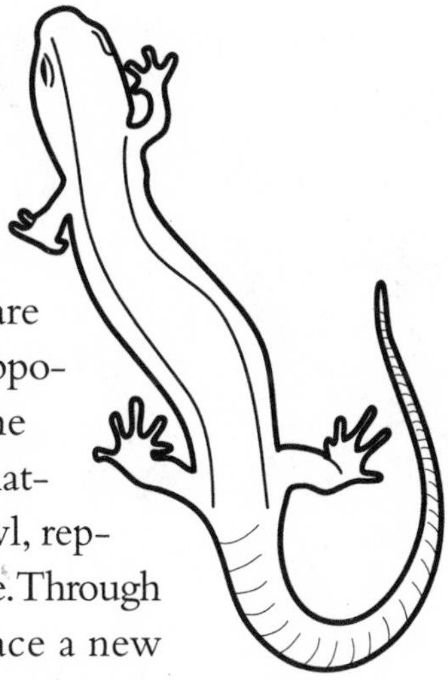

and propelled backward along the floor. (Fig. 4.6) This whole side
of the body, from hand to foot, is now elongated and bears the
weight, allowing for mobility in the opposite side which has flexed
during the action. This causes the spine to curve laterally, parallel to the floor in the vertical plane of movement (the planes
are defined relative to the moving body, so are not fixed
in space). The Homolateral pattern is seen in many reptiles. (Fig. 4.7)

As the opposite leg flexes in, it prepares to push;
the impulse from the foot pushing against the floor
travels upward along the spine and through the arm
on the same side. The leg, torso, and arm of this side are
now fully elongated and bearing the weight, while the opposite side flexes in preparation for another push from the
foot and foreleg of this side. (Fig. 4.8) In this way, alternating the initiation from side to side, the infant can crawl, rep-
tile-like, forward and back, and now becomes very mobile. Through
this pattern the child can also turn itself around to face a new
direction and gradually, as spinal and lower limb strength and coordination develop, push back into sitting and then forward onto
the hands and knees. (Fig. 4.9)

Each transition from one level to another is simultaneously
initiated by a reach and pull through the spine and supported by

*Figure 4.7
The Homolateral Push
patterns are seen in
many reptiles.*

77

Figure 4.8 The Homolateral Push from the lower extremities.

one of the push patterns. Either the push or the reach and pull may tend to dominate the way the infant initiates its movements.

Either tendency can be, but is not necessarily, a sign of incomplete development of the other phase; it may also reflect an innate tendency in the personality to prefer a particular quality of attention and expression. We do not need to try to change this preference, but simply ensure that both patterns are available to the individual, enabling her to find her own dynamic balance of interaction between initiation and support for movement.

Figure 4.9 The Homolateral pattern is used for support as the tail of the spine pulls back into sitting.

The Contralateral Reach and Pull Patterns

The last and most sophisticated of the developmental patterns is the Contralateral Reach and Pull, again initiated first from the hands, which pull the body forward through space, then from the feet pulling backward. This pattern begins to appear when the infant is able to support its body weight on hands and knees. The infant will probably have first gone through a short phase of moving about in this posture either

Figure 4.10
The Contralateral
Reach and Pull from
the upper extremities.

through the Homologous or the Homolateral Push patterns, or both. Once a certain amount of stability has been achieved, attention will again be drawn out into the environment, and the desire to reach out toward a friendly face or a toy with which to play will activate the intent and the will to move beyond the security of its known boundaries. The hand, coordinated with the looking of the eye that expresses intention, will reach forward and pull through the spine and the leg on the opposite side of the body. Weight then shifts to a diagonal support and the other hand will reach forward, pulling through the opposite leg. (Fig. 4.10)

In the same way, movement backward is initiated by the reaching of the foot. When the attention is alert and focused outward, the tips of the toes initiate, with the vision supporting. Here the child may be retreating from the object of stimulation or moving toward something behind it. This pattern is normally used to turn around and face behind, and to change levels. In crawling backward we are also learning to measure, or gain a "felt sense" of, the space behind us. (Fig. 4.11)

A sequential rotation through the spine underlies the action of the Contralateral patterns and creates the possibility for movement through all planes simultaneously. Such movement is a spiraling and allows for continuous transitioning between levels and directions. The Contralateral pattern has a distinctly different

Figure 4.11
The Contralateral
Reach and Pull from
the lower extremities.

Figure 4.12 Most
mammals display the
Contralateral pattern in
their normal walking
gait. Although at certain
phases of the movement
the coordination appears
to be the same as for the
Homolateral pattern, the
initiation is different.

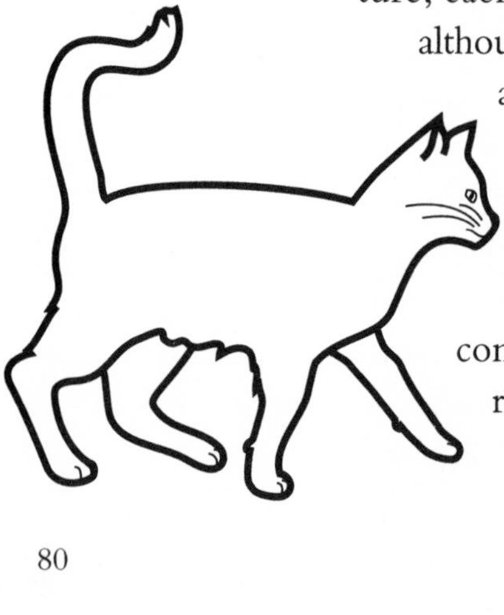

rhythm from the Homolateral, much lighter and often more swift and fluid. It may express the quality of a wild cat stalking its prey, alert and sensitive. This is a very active and lively phase of growth, and the child will display great energy and determination in its efforts to master the new patterns as it explores the ever enlarging world now made accessible to it.

The Contralateral pattern, commonly known as "creeping" or "cross-crawling," is seen in the walking gait of most mammals. (Fig. 4.12) It is also the basic pattern of our bipedal walking, running, and so on. Through an integration of all previous patterns, the child learns to push and pull itself up into the vertical posture; each child will find its own unique way of doing this, although the underlying principles are the same. The Brachiation pattern of the primates—apes, chimpanzees, and monkeys—is an intermediate phase between the horizontal and vertical stance. In this transition, the hands reach for and grasp onto a support and help to pull the body upright. (Fig. 4.13) The child uses this pattern to come to standing, supported by a push from the feet and reach of the head. Once standing, she will still use her hands to support for quite some time while mastering balance. She will often take her early steps while clutching onto a favorite toy, a parent's hand, reaching out to touch a wall or chair, or simply reaching out against space. The child is using the kinesthetic memory of the sensation of the ground's support beneath her hand to find support in this new and precarious relationship to gravity.

It will take most children a little more than the first year of their life to reach this stage. It then takes many more years to fully master and integrate each pattern into the vertical posture and to explore the endless possibilities of their expression in both functional and creative activity. In all the movements of rolling, crawling, and creeping, the spine is primarily in a horizontal relationship to the ground and thus more securely supported. These movements are an essential preparation for the Contralateral patterns. In the early patterns the spine and the extremities of the body are developing an integrated relationship to each other and to gravity through movement in the horizontal plane. The way the body moves creates its postural attitudes or alignment. These attitudes and movement patterns, developed in infancy, are a foundation for and will be reflected in the vertical posture and movements of the child and adult.

Figure 4.13
The Brachiation pattern is seen in the primates.

The spine needs to feel itself supported, initially by direct contact of the torso with the ground. Then the support comes through the limbs, first pushing against the floor and then extending into the environment and surrounding space. To the extent that this support occurs and is integrated through movement initiated peripherally, the spine will have freedom and mobility. If the support of the limbs is lacking or incomplete, the spine will have to support itself once it is vertical. This is done through holding the body centrally, which creates a pattern of tension and rigidity in the spine and surrounding tissues and organs. In this particular pattern the limbs usually lack a sense of active presence and aliveness. There is weak or flaccid muscle tone in them, and the hands, feet, and senses are not in direct and active contact with the external environment. Attention tends to be inner-directed and may not shift readily in response to life's changes; rather it is statically held around an inadequately supported center.

The Spiral

We have seen the way in which the sequence of patterns unfolds in overlapping waves. This is neither a linear nor cyclical process. Although there is a similarity in the process of change experienced at each stage and transition, there is also a movement upward, a change of level at each point in the turn of the spiral. In the developing movement of the infant, the reaching beyond personal boundaries initiates the upward pull; the intent and purpose behind this action is created by a shift of attention outward and upward and by the desire to move beyond what is known. Circling at any level of the spiral represents the necessary period of integration and mastery of each new level and preparation for the next transition.

Exploration: The Sequence of Developmental Patterns

Working with the developmental patterns helps to clarify neuromuscular coordination, perception, and attention, and strengthen the foundations of movement; each pattern provides a base of support for the next stage. As you practice the movements outlined below, note which patterns you enjoy, dislike, avoid, or find more difficult. Your experiences will reflect your personal strengths and weaknesses within the developmental process and can be a key as to where you could benefit from further practice. If a particular pattern is not being expressed in movement, perhaps due to incomplete development in childhood or loss of expression through a limited range of activity in later life, its potential can be reawakened through practice.

Table 1 gives the order in which the patterns emerge to use as a reminder of the sequence. You can explore each stage separately by practicing the movement exercises described here, which may be seen as a crystallization of each stage; or, using the principles outlined to guide you, find a variety of movements which fall within each basic patterning. Explore each one in different positions and relationships to gravity: lying on your back, front, or side; sitting; kneeling; on hands and knees or feet; or standing. You might also like to explore the patterns by imagining yourself as a fish, reptile, or similar creature in order to discover the movement patterns, qualities, and mind states of each in an original and creative way. This can open up a whole new world of creative movement possibilities.

Alternatively, you might like to begin with the Cellular Breathing and Navel Radiation patterns. From here improvise your own way through the developmental process in a spirit of exploration and play. If you notice you tend to avoid or have difficulty with a certain pattern, you can go back to practice that one in a more specific way later.

See if you can feel how one stage underlies and supports the next, how yielding and pushing transforms naturally into reaching and pulling, and how the transitions from one pattern to another are made.

TABLE 1. DEVELOPMENTAL PATTERNS

Pattern	Body Coordination, Species, Age
1. Cellular Breathing	Expansion and contraction of each cell of the body in internal respiration. Integrates and aligns the physical body. The original one-cell (ovum). One-celled organisms, e.g. amoeba. Present throughout life from conception, as it underlies breathing and all life processes. Mind of "being" predominates.
2. Navel Radiation	Integrates the extremities of the body into the center, through the navel. Starfish. Present in utero.
3. Mouthing	The head rocks on the lower jaw; nursing action of the infant. Hydra, sea squirt. Prebirth and birth; dominant during early infancy.
4. Pre-Spinal	Integrated movement between head and torso to the tail; spinal movements initiated in the "soft spine" of the spinal cord or organs. Underlies spinal patterns. Lancelet amphioxus. Pre-birth, birth, and early infancy. Transition to "doing" mind.
5. Spinal Push from Head 6. Spinal Push from Tail	Integration of the spine from head to tail; spinal movements, initiated in the musculoskeletal structure. Inchworm, caterpillar. (See p. 52 and note 6, p. 312.) Pre-birth, birth, early infancy.
7. Spinal Reach and Pull from Head 8. Spinal Reach and Pull from Tail	Movement of the spine through space, led by the head or tail; enables the child to change levels. Fish. Birth, early infancy, with initiation first from the mouth; other senses develop in the first few months.

TABLE 1. DEVELOPMENTAL PATTERNS *(Continued)*

Pattern	*Body Coordination, Species, Age*
9. Homologous Push from Upper Extremities 10. Homologous Push from Lower Extremities	Both arms and hands together push the body backwards; then both feet and knees push the body forwards. Amphibians; rabbit, kangaroo, other mammals, e.g. horse, dog, when running at speed. From upper: birth to three months. From lower: three to five months.
11. Homologous Reach and Pull from Upper Extremities 12. Homologous Reach and Pull from Lower Extremities	Both arms reach forward and pull the body through the space in front; then both legs reach backward and pull the body through the space behind. Fingers and toes initiate. Frog leaping, squirrel; other mammals when running at speed. Five to seven months.
13. Homolateral Push from Upper Extremities 14. Homolateral Push from Lower Extremities	Belly-crawling: the push from the right arm sequences back into the right leg, elongating the right side and flexing the left. This prepares for the push from left foot through to left hand, and for forward movement. Alternate sides initiate; the initiating side elongates. Amphibians and reptiles, e.g. lizard, alligator. Some mammals, e.g. camel, elephant. Others revert to this pattern when trotting at a moderate pace. From upper: five to six months. From lower: six to eight months.
15. Contralateral Reach and Pull from Upper Extremities 16. Contralateral Reach and Pull from Lower Extremities	Cross-crawling on hands and knees, walking, running, etc. Fingers of one hand reach forward to pull the opposite leg through and move forward. Toes reach backward to pull the opposite arm through and creep backward. Most mammals when walking. Humans. The Brachiation pattern of primates (e.g. apes) involves a reach and pull through the hands. From upper: seven to nine months. From lower: nine to eleven months. From about one year onwards. Contralateral walking and running gradually develops.

All patterns, once developed and integrated, continue to be refined and strengthened throughout childhood and adult life. They are, under normal circumstances, present throughout the whole of life. Ages given are approximate times a pattern fully emerges and temporarily dominates. Individuals may vary greatly in this timing, and also in the actual movements they make within each basic patterning. The natural order of the unfolding of the sequence of patterns appears to be universal, but usually occurs in overlapping waves of development unique to each child.

When all of the patterns are fully integrated we can spiral fluidly through changes in levels and directions.

Movement Repatterning

Chapter Five

Developmental Movement Therapy

The notion of perfection has no meaning in relation to natural phenomena. Nature's process includes events of disorder, chaos, decay, and death. These experiences often defy the intellect's attempts to make sense of them by labeling them; they may be labeled as "failures." Hence, when we talk of the natural development of a child or human being, we must assume that the process will be imperfect, that sometimes the plan will go wrong and areas of so-called failure or weakness emerge. These may result from conditions in either the inner or outer environment of the child or adult. More often, though, they arise in the interface between the two worlds—where the young infant's individuality begins to develop and interact with the unique environment into which it is born.

This is as true for the development of movement in the child as it is for its psychological growth. Each of us may retain into adulthood some areas of weakness in our movement ability resulting from early learning difficulties. For the majority of us, these "failures" are not so profound as to significantly disrupt a relatively normal process of development. We are able to accommodate or compensate for such gaps and still function in a perfectly adequate way, yet some area of our potential remains unfulfilled. For a child with more severe learning difficulties, however, the whole process of movement, perceptual, psychological, and intellectual development can be seriously impaired.

The causes of such problems are varied. They include damage to the brain and nervous system before, during, or after birth, or abnormalities in their development in utero; hereditary or congenital illness; inadequate meeting of the basic survival needs such as proper nutrition, warmth, and a healthy environment; lack of physical holding when very young and appropriate parental encouragement to move and learn; and suffering of emotional trauma without the subsequent support of a loving holding environment to contain the overwhelming feelings. Emotional distress or physical illness in either the child or its close family may temporarily inhibit its ability to fully practice and develop the movement patterns emerging at that time. There may also be simple but often unrecognized physical factors that make certain movements uncomfortable or impossible to perform, such as floor surfaces or clothing which can restrict movement. For some infants, their parents' ambition to see them walking early, perhaps together with their own either compliant or adventurous nature, may encourage them to "grow up" too quickly, missing out on much valuable learning experience in the early stages of development.

If nature is able to create disorder, chaos, and "failure," she is also able to recreate harmony, wholeness, and "success" from them. In most of the situations described above the unfulfilled potential is not completely lost; it is merely not yet experienced, or it has been experienced and then forgotten. It is possible for us even as adults to remember these potentials for movement experience, thus clarifying and strengthening the foundations upon which our present movement, perceptual, psychological, intellectual, and spiritual experiences are based. This is the process to which Developmental Movement Therapy addresses itself. We find that this reeducation of underlying developmental patterns not only gives more inner strength, clarity, and aliveness to our movement and perceptual responses, but it also frees more of our energy for creative thought and activity.

The sequence of movement patterns will unfold according to its own innate timing; regardless of whether a pattern has been embodied or not, the next pattern will emerge at its appropriate time. If a pattern has been missed, this gap will weaken the support for all subsequent movement development. Generally, we learn to compensate for such weaknesses, usually through the creation of rigid holding patterns, from which areas of tension and immobility in the body ensue. Much energy is used up in maintaining such an inefficient method of support, and that energy ceases to be available for creative work and play.

For example, suppose a child has been seriously ill for the second half of its first year and is unable to move around as a child of its age would normally do. At eighteen months this child may be able to walk adequately well for its age but may be holding tension in the spine, neck, shoulders, or pelvis to do so. If it has not had sufficient practice supporting the spine on all fours, support will be maintained through holding centrally; this is the tendency described in the previous chapter. Such a child may also have had too little practice in making the gradual transitions from lying to standing and so doesn't know how to safely return to the floor by itself. This can produce a lot of anxiety and fear when standing, which will be reflected in the tension and rigidity of the central holding pattern. Such compensation need not stop the child, and adult, from performing a wide range of movement skills, but it will inhibit full expression and enjoyment.

Many people in our goal- and thinking-oriented society have forsaken taking real pleasure and satisfaction in the pure beauty and vitality of movement. This pleasure begins in early infancy, and will be affected by the attitudes of the family and culture in which the child is brought up. If the culture says that mind and body are separate, the child will soon learn not only to differentiate them, an essential process in psychological development, but will also unnaturally separate them. Similarly, if the culture conveys the message that the body is in some way degraded, sinful, or inferior to the mind, the child will learn not only to control and moderate the impulses of the body, another natural process in healthy development, but also to repress the vital energy that is a human birthright. However, the potential for wholeness remains intact and can be rediscovered. Recreating patterns of movement that express the free and full flow of energy throughout the body is one means of reclaiming our body-mind birthright.

Neurological Connections

Each basic developmental movement pattern is coordinated by a specific area of the brain. Messages are sent through particular pathways of the central and peripheral nervous system to the muscles that will perform that movement sequence. This normally happens automatically, stimulated by the desire and intention to make the movement. The movement is coordinated by the brain and central nervous system with sensory information from other areas of the body and from the environment as perceived through the senses. Movement is performed with greatest ease and clarity when the most direct nerve pathway to and from the brain area is activated. Damage to or dysfunction in the appropriate brain area or nerve pathways will inhibit the related movement pattern.

In many cases it is possible to facilitate the performance of an inhibited movement pattern through restimulation of dysfunctioning areas of the brain, or through what we might think of as a rerouting of the messages through nearby undamaged cells. What we feel when we contact the brain tissue at a cellular level is movement and stillness. We feel areas that are active and through which movement can pass freely, overactive and "overcharged" places, and areas where there is a quality of inactivity, stuckness, or darkness. It is in these "dead" areas that we feel function to be inhibited through either a lack of aliveness and cellular awareness in that area or damage to the cells. In the first case, awareness can be awakened through contact and movement at the cellular level. In the case of damage, we would also work with the sense of rerouting, or creating new neural pathways, as movement innovator Emilie Conrad-Da'Oud has also done with those who have sustained a spinal cord injury.[1] It must be noted that this way of working is based on experience and observation rather than scientific findings. Although the brain has been fairly extensively mapped, there is much that is still unknown about its workings and how the healing of dysfunction or damage to the brain actually takes place. It may be that some cells are able to alter their functions as needed; if so, perhaps these cells can take up the functions of damaged cells, thus creating a new route for the movement impulse through the nervous system.

Conversely, areas of the brain and nervous system can be stimulated through the practice of specific movements. It is the actual practicing of the movement patterns that stimulates the developing functions of the brain areas involved, both in the normal course of development and in the application of Developmental Movement Therapy.[2] This also gives support to the other physiological functions for which that brain area is responsible. When higher areas of the brain have had to take over the functioning of the low brain because it has suffered damage, the natural processes of these higher areas may be severely restricted and underdeveloped as a result. Without the underlying support of automatic coordination of movement and bodily processes by the lower brain areas, a child with severe low brain damage will not be free to develop its creative, intellectual, and social potential fully. (Fig. 5.1)

Table 2 (pages 96–97) indicates relationships between the developmental patterns and areas of the brain primarily responsible for their coordination. Included is a simplified outline of the functions associated with each brain area. This information is adapted from writings of Bonnie Bainbridge Cohen, including a work-in-progress and unpublished notes. It should be looked upon not as a fixed formula but as an attempt to give form to observations made in an ongoing exploration process.[3] What is

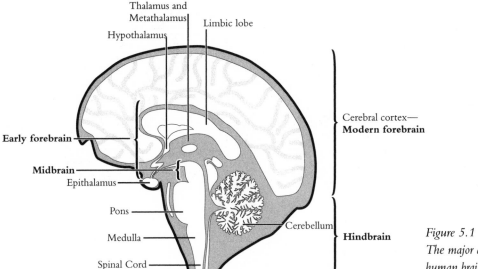

Thalamus and Metathalamus
Limbic lobe
Hypothalamus

Cerebral cortex—
Modern forebrain

Early forebrain

Midbrain
Epithalamus

Pons

Medulla

Cerebellum

Spinal Cord

Hindbrain

Figure 5.1
The major areas of the human brain.

offered here is, in Bonnie Bainbridge Cohen's own words, "a working model from which to begin and [from which] to measure and reevaluate." She emphasizes that as what we are observing in our research is the fully functioning person in relationship to community and environment, the variables are naturally enormous.

Each of the developmental patterns is also given energetic support by a specific gland of the endocrine system. I include this information here for reference, but the reader is advised to refer to the section on the endocrine system in Chapter Eight for a discussion of these relationships and the inclusion of certain non-endocrine structures (called "bodies") within the endocrine system.[4]

TABLE 2. DEVELOPMENTAL PATTERNS

Pattern	Endocrine Glands	Brain Area
1. Cellular Breathing	(Each cell)	(Each cell)
2. Navel Radiation	Adrenals	"Abdominal brain," Solar Plexus
3. Mouthing	Endocrine cells in Small Intestine	Spinal cord and Lower Medulla
4. Pre-Spinal	Thoraco Body	Spinal cord and Upper Medulla
5. Spinal Push (Head)	Pineal	Hindbrain: Lower Medulla
6. Spinal Push (Tail)	Carotid Bodies	Hindbrain: Upper Medulla
7. Spinal Reach and Pull (Head)	Mamillary Bodies	Midbrain: Lower Cerebral Peduncles
8. Spinal Reach and Pull (Tail)	Pituitary	Midbrain: Upper Cerebral Peduncles
9. Homologous Push (Upper)	Heart Bodies	Midbrain: Inferior Colliculi
10. Homologous Push (Lower)	Pancreas	Midbrain: Superior Colliculi
11. Homologous Reach and Pull (Upper)	Thymus	Ancient Forebrain: Hypothalamus
12. Homologous Reach and Pull (Lower)	Thyroid	Ancient Forebrain: Thalamus
13. Homolateral Push (Upper)	Gonad of same side	Hindbrain: Cerebellum/Pons
14. Homolateral Push (Lower)	Coccygeal Body	Hindbrain: Cerebellum/Pons (Anterior portion)
15. Contralateral Reach and Pull (Upper)	Opposite Inferior Parathyroids	Modern Forebrain: two hemispheres of Cerebral Cortex
16. Contralateral Reach and Pull (Lower)	Opposite Superior Parathyroids	

Associated Functions of Brain Areas	*Evolutionary Level of Patterns*
1. Basic life processes; internal respiration and energy production; "being" nature.	Invertebrate: One-celled organisms (e.g. amoeba). Water dwelling.
2. Integration of the limbs into the center; digestion and basic metabolic processes.	Invertebrate: Echinoderms (e.g. starfish). Water dwelling.
3. Control of vital visceral functions, e.g. breathing, digestion, circulation by Medulla; establishing of vertical axis with Pons and Midbrain; Spinal Cord relays sensory and motor information to and from brain and controls some primitive reflexes.	Invertebrate: Tunicate (e.g. sea squirt). Water dwelling.
4. Birth of "doing" nature. As for 3.	Invertebrate: Lancelet amphioxus. Water dwelling.
5 & 6. As for 3.	Vertebrate. Inchworm. Land dwelling. (While it is not a vertebrate animal, the inchworm is used as an example of this movement pattern. See page 52, and note 7, page 314.
7 & 8. Major motor pathways between Forebrain and Hindbrain.	Vertebrate: Fish. Water dwelling.
9 & 10. Relay centers for auditory and visual impulses.	Vertebrate: Amphibian. Water and land dwelling.
11 & 12. Hypothalamus is important in many emotional and visceral processes; control center for Autonomic Nervous System, so maintenance of homeostasis. Thalamus is a relay center for sensory information; regulates pleasure/pain reflexes. Nervous and Endocrine systems meet in Ancient Forebrain; seat of perception.	Vertebrate: Amphibian. Water and land dwelling.
13 & 14. With Medulla and Midbrain, Pons helps establish vertical axis; Cerebellum and Pons control balance around and falling off vertical axis; Cerebellum is center for automatic coordination and control of movement.	Vertebrate: Reptile. Land dwelling.
15 & 16. Center of integration of many complex motor and perceptual functions; the Cerebral Cortex is the seat of conscious learning, intelligence, creative and intellectual thought, imagination, and communication and language skills.	Vertebrate: Mammal. Land and tree dwelling.

The Art of Movement Repatterning

The principles of Developmental Movement Therapy can be used as a diagnostic aid, through observing how the movement patterns are expressed or not expressed by the infant, child, or adult. The neuromuscular system can then be reeducated appropriately to stimulate, strengthen, or redefine patterns that are absent or difficult to perform clearly.

The practitioner must have a clear vision of the pattern that is not being fully expressed and must be able to hold this as a context for the work she does. This also entails her knowing the pattern clearly in her own movement experience, so a practitioner's training always involves in-depth personal experience of the material. As long as the potential for a movement exists, it may be seen as the unexpressed or shadow side of what is actually visible. By holding this vision in our awareness, we allow for the possibility of the student's movement to align itself with the perceived potential. We choose not to limit our focus only to the problem or difficulty presented. While it is at times appropriate to give specific attention to the problem, if we don't hold it within the larger vision of possibility we may find that we become stuck in self-defeating circles. We then risk giving more energy to the area of pathology, through the continued direct focus of our own and the student's attention. The problem must be seen within the context of the whole, which includes the inherent developmental pattern and its place in the overall development of this individual.

If we are able to hold within our awareness the movement potential we are seeking to help free, then we create a space within which the movement pattern might be embodied and experienced by the student. There is recognition of the pattern within the nervous system at the moment when the movement is fully integrated, a moment of clarity when the student feels without

Tension!
link to awareness

doubt fully present and at ease in the movement. This moment of recognition may or may not be available to the conscious mind. It will most certainly be experienced as a change in energy, attention, and perception as the pattern is discovered, or rediscovered, and new movement sensations are felt. First we simply *feel* the new sensations, perceptions, and mobility in a very direct and immediate way, as does an infant or child. Conscious recognition of change may occur in the adult only after the movement has been fully experienced; integration at the cortical level of the brain is not necessarily simultaneous with the change in movement patterning.

The very first moment of recognition is the most significant one in the process of repatterning. It is then that the greatest learning within the neuromuscular system is taking place: a new connection is being made and a new alignment of mind and body is discovered. Within each of us there is an innate tendency towards health, growth, and the fulfillment of potential. The body-mind will usually, if allowed, choose the most natural, healthful, and efficient patterns available to further this tendency towards wholeness. Once a new, more whole and efficient pattern has been experienced and recognized at some level of awareness, that pattern becomes potentially available to us. Instead of being stuck in an habitual or limited pattern of movement, perception, and behavior, we now have a choice. The new pattern needs to be supported by further practice and refinement and nurtured by understanding, encouragement, and appreciation of the effort involved. The student will then be more and more able to make her own choice toward healthy functioning.

In performing the Developmental patterns we want to see an unobstructed flow of energy through the body throughout the *initiation, sequencing,* and *completion* of the movement. A clear movement pattern initiates at one or more of the extremities and pulls or levers directly through the center of all joints of the limbs and spine to the opposite extremities, where it completes itself.

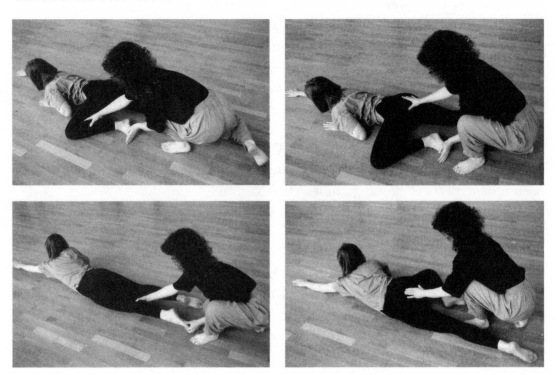

Figure 5.2
In movement repattern-
ing attention is paid to
the initiation, sequenc-
ing, and completion of a
movement.

(Fig. 5.2) We are also looking for an alignment of *attention* (being present to what is), *intention* (the subtle movement of mind that precedes action), and *activity* (the performing of the intended action). (Fig. 5.3) Some of the reasons why such choices of patterning may be unavailable to the individual have been outlined above.

It is important for the educator or practitioner to observe without judgment and to respect where each person is now in her development. Our own need or desire to see change and positive results immediately can interfere with our student's own natural process of learning or healing. Obstructions to the free flow of movement may have developed out of an early need to create defense or stability, when the child was unable to find more effective means of protection or support of its own accord or through its environment. We need to be sensitive to the student or client's degree of readiness to let go of old patterns. What we are offering is an alternative way of developing inner strength and balance, but if at some level the person does not feel ready to make the change it is not our place to attempt to go against their will. When there is genuine resistance there is something still to be learned or discovered in the place where one is stuck, and we cannot move on until that learning has taken place.

Figure 5.3
The alignment of attention, intention, and action.

The function of the practitioner is not to manipulate the body passively into a new pattern of movement, but to provide support and guidance for each person, whether adult or child, to make her own discovery in her own time. In this way the "mind" of the movement changes and there will be a true and lasting learning, not merely a mechanical imitation that is perhaps dependent on the practitioner's continual presence for guidance.

With our own mind we make contact with the mind of the student, to know where in the body the mind and energy is moving or obstructed. The mind directs the energy and this creates bodily movement. Through the touch of our hands, the most sensitive and articulate parts, and through our own sensory perceptions, we are able to feel this movement and stillness—its quality, direction, and degree of freedom. As our mind settles to a state of quiet receptivity, we become attuned to the subtle stirrings of an intention in our student to move, perhaps before she herself is aware of it. This very subtle level of movement initiation is going on continually throughout the body, usually below the threshold of consciousness. Such movements are in response to a myriad of thoughts, images, feelings, bodily processes, sensations, and sensory perceptions, all of which are constantly arising and changing. It is at this level that our habitual movement patterns first begin to be established, as unconscious responses to our inner and outer environment initiate or obstruct the movement of energy through the body in a particular way. We will return most readily and automatically to the patterns which are most consistently reinforced, particularly when we are under stress. In this way the patterns the mind creates may either contribute to the healthy functioning of the whole person, or to a state of nonintegration and imbalance.

In seeking to introduce an alternative patterning, we work at this subtle level of mind meeting mind, attuning ourselves to movement initiation that is usually below the level of conscious awareness. Through our hands we receive information about where the student is able and chooses to go, where she is at the present time unable or unwilling to go, and where she may wish to go but cannot because she is caught in some other habitual groove. Through our hands we also convey to her the vision we are holding of the movement potential that we may help to free. We feed in the direction of this movement, waiting until it meets with recognition and the willingness of the student's mind to initiate this pattern. Again this may be at an unconscious level, but if we can coordinate the timing of our direction with her subtle intention to move, felt through our hands, we can guide the student into the new patterning.

Depending on the nature of the origins of the inhibition, the full sequencing and completion of the pattern may be found immediately once connection has been made to the initiation. In other cases the process may require more time and further guidance to integrate the whole body into the movement sequence. However, no matter what the original cause of the obstruction, the pattern will not emerge clearly and fully until the initiation is found and coordinated actively and willingly into the movement pattern. The beauty lies in the fact that no matter how severe the damage or inhibition may be, the intention to initiate these movements seems always to be present; we need only know how and where to look for it. This intention connects each person with their will for life, and so to their self-healing process.

From what has been said in previous chapters it will be seen that clear initiation of the Developmental patterns depends on the alignment and integration of inner sensations, response, and willingness, with stimulus from the environment in the form of the supportive pull of earth's gravity along with sensory perceptions received through the skin, mouth, nose, ears, and eyes. As well as using our hands to feed in and receive information about bodily movement and sensation, we will also provide or utilize already present external stimulation appropriate to the movement pattern being practiced. This may take the form of having the student position her body in such a way as to experience more clearly the fall of weight into gravity through certain parts of the body, such as the head, elbows, forelegs, and front or sides of the torso. We may also feed in a gentle compressive force with our hands through a specific body part to simulate the meeting of the force of gravity with her body's weight. This can be done to facilitate a pushing action through that limb or body part, or to stimulate the feeling of connection to other parts through proprioceptive feedback. (Fig. 5.4)

Figure 5.4   A gentle compressive force is fed in though the legs; this simulates the falling of weight through the bones and joints, and facilitates a pushing action from the feet through the whole leg.

Figure 5.5 Stimulation of the senses to facilitate reaching out.

Touch to the skin of the face, hands, or feet or the presence of scents, sounds, and objects in the environment may be used to initiate reaching actions. In coordinating sensory and motor activities, the student is guided into reaching out towards the source of stimulation. (Fig. 5.5) If the student, either as an infant or adult, is still at a very early level of development, the most direct reach may be that of the mouth toward food. In such a case a favorite food might be used to stimulate the initiation from the mouth of the early reaching movements when no other responses are available. Once initiation and performance of the earliest patterns have become easy and familiar, we can then proceed through subsequent stages to the limit of the individual student's capability.

In cases in which even the rooting reflex and active reaching of the mouth are absent or when touch to the face and head or hands is resisted, we may need to return to a prior level of development. Physiological flexion and extension develop in utero and provide an underlying quality of muscle tone throughout the whole body, in preparation for actual muscular activity. They underlie the development of reflexes. In order to balance the body around its vertical axis this tone needs to be developed equally along the front and back of the spine and along each side of the joints of the limbs. If flexor tone on the front of the body is very low, for example, the muscles there will not be able to counter the strong pulling of the back extensor muscles as they develop. The body will then be thrown into an exaggerated backwards curve, as seen in some spastic conditions. A lack of development of physiological flexion in utero may be one underlying cause for this condition. Or the whole body musculature may suffer from either too high or too low tone, reflecting an overall pattern of over- or underdevelopment of both physiological flexion and extension. Unlike individuated flexion and extension (which, as we have seen, develop from the head downward), physiological flexion and extension develop in utero from the feet up.

Tactile stimulation to the feet and forelegs can help awaken neuromuscular response and coordination in those with severe movement disability or developmental delay. This touch to the feet is usually received more willingly and more often with enjoyment than touch to the face. Simple stroking and tickling of the feet or light skin massage can be used and may be a first way to make contact with someone whose mind has been otherwise unreachable. Involuntary movement of the muscles in response to this light touch is continually giving sensory feedback to the brain, and gradually awareness is awakened in the body as muscular tone increases. Unpredictability in the quality, speed, and rhythm of such stimulation is necessary to keep awareness in the present moment; it enlivens the nervous system and reflects the unpredictable nature of life itself. To a normally functioning person, this kind of stimulation to the feet can bring an alertness

*Figure 5.6
Awakening sensation in the feet gives them a sense of active grounding and aliveness.*

of attention and a deeper sensation of connectedness through the soles of the feet to the ground. (Fig. 5.6)

In working with someone it is important to start from the level at which she feels comfortable and secure. If there is a problem with a particular stage of movement development—recognized as a pattern that is absent or difficult to perform

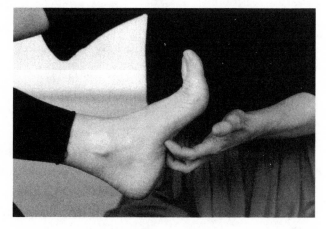

or a stuckness at a certain stage so that all movement is tending to reflect the coordination, quality, and "mind" of that stage and not develop further—then we need to allow the student to work first with what is easy and comfortable for her. This may mean going back to an underlying stage of development prior to the difficulty, or allowing her to be where she is stuck—there may be some important, yet to be discovered learning within this pattern that is keeping her in that place. The wisdom of the body understands this need and will tell us of it. We need only to listen to its messages.

We are looking to facilitate not only the performance of certain movement sequences in repatterning, but also the transitioning from one level of movement development to the next. It is in the infant's actual experience of making the transitions that it learns, grows, and develops its sense of individuality and competence in the world. The learning is then not something that is given to the infant but is its own achievement, and each experience inspires strength, confidence, and joy. This is just as true for adults. If we know how we took our first step into the new territory, with all its hazards, joys, and efforts, and if we really know that we made that step ourselves, then we have much more strength, courage, and trust in ourselves to take the next step. In working with a student we play between giving support to whatever is happening and holding the vision of what is possible. Within this play we create a space in which the transition or transformation may take place.

Body Memory

In working at this deep level of the body—going to the source of movement and bringing awareness to its patterning at a fundamental cellular level—emotional memories, feelings, and associations may also be brought to consciousness. Memories of past experiences are stored within the tissues and fluids of the body;

these are often lost to consciousness as the mind, unable to integrate them at the time, numbs itself to their presence by blocking the free flow of energy and movement through the area. Such stagnation seems to occur mainly in the body's fluids and soft tissues. Reawakening of awareness and movement there may cause memories and feelings to resurface into consciousness.

In utero and early infancy the mind and body, psyche and soma, are not yet differentiated but are still experienced as one. Until body and ego boundaries are established, the inner and outer environment are also experienced as undifferentiated. If we touch upon a pattern of blockage or disconnection rooted in the experiences of very early life, the freeing of energy flow there may carry with it the release of associated emotional energy. Such events were experienced originally through the fetus' or infant's preverbal consciousness, before it possessed words or concepts with which to discriminate experience. Therefore, this related event may not stimulate direct mental associations. Instead the experiences may be perceived as bodily sensations, feelings, or images.

Responses to this work are often gentle and may include the deepening of the breath, a general sense of relaxation or well-being, or a change in perception and awareness. Sometimes less comfortable bodily sensations, such as pain, nausea, or deep tiredness, may be experienced, or emotional expression may occur. These responses are natural to the process; they correspond to the reawakening of sensation in areas of the body that have been numbed, or to the pain and frustration of being blocked. The release frees or discharges the emotional content that has been withheld, and discomfort is often a sign that the healing process has begun. There can be no movement toward the healing of old wounds as long as we remain unconscious and numb to our pain and fear and disassociated from imprisoned areas of the body.

However, we should never push the student beyond her own ability to contain and process feelings or her own willingness to proceed. The body-mind in its own wisdom knows how much change it is ready to receive. We can learn to discern whether an emotional response is a sign of genuine need to resist going further, or an expression of the pain, fear, or frustration that is often inherent in the process of learning and transformation. In the latter situation, the quality of response is recognizably different, and we feel the student to be in cooperation with us and willing to accept the challenge. It is often helpful to obtain verbal consent as well; this makes explicit the unspoken, strengthens trust, and supports the student to be self-directing. With an infant or child we must, and can more easily, rely on our senses to discern when "no" means "yes," and when it truly means "no."

In working with an adult it may be necessary to help integrate material with strong emotional content through verbal processing and/or some form of creative expression, such as drawing, dancing, writing, or singing. Including these processes in a session can always bring greater insight and self-awareness and will help the student more deeply integrate and "own" what has transpired. Dialogue is particularly helpful if associations to experiences from later phases of life, when mental and verbal abilities have developed, arise through the work. However, we should also be aware that preverbal experiences engender basic patterns and attitudes that recur throughout adult life. Later experiences will tend to reflect and re-present the earlier ones that seek resolution or completion. The degree of both emotional response and need for conscious verbal processing seems to depend to some extent on the individual's unique way of organizing and experiencing the relationship between her body, mind, and feelings.

Generally the emotional aspect of the process is well contained, as we are not only releasing the hold of old patterns but simultaneously introducing new, more helpful and flexible ones. In this work deeper sources of inner strength, core support, and integration are accessed, empowering the student or client gradually in the process of movement repatterning. These new patterns may feel strange at first: unfamiliar sensations are now being received from the body and as a result the brain is reorganized. But as the sensations become familiar and are integrated into everyday movement activities, new qualities of strength, openness, coordination, and grace are found. The expression itself through movement or other creative activity is a means of simultaneously containing and processing emotional content.

We not only repress painful and unpleasant feelings and sensations but also the essential joy, vitality, harmony, power, and love that are our true nature. It is likely that all of these qualities have been rejected or disallowed at times in our lives. They will begin to reemerge from deep in the tissues of our physical bodies as we open to more and more of our wholeness.

The Healing Relationship

The practitioner's simple presence gives the student or client another person to be touched and heard and seen by, to listen to and move toward or away from, to meet and be met by. The practitioner is the essential bridge, the human link that will enable the student to take a step into the unknown and do something never done before. At any kind of transition point in our lives we all need the certainty of knowing at some level that we are securely held. This is the ground from which we leap, however near or far, and to which we can return and be welcomed back. We also need the experience of being witnessed in our moments of pain and joy, growth and change. The infant needs to feel its mother's presence holding it as it makes its first gestures of assertion and independence: the "I am" of its existence. So too does the student or client need to feel the supportive presence of someone who can be trusted in order for any new learning or experience of self to take place. "At such moments the infant relies for safety on the holding environment. 'Only if someone has her arms around the infant at this time can the I AM moment be endured, or rather, risked.'"[5]

And as our steps take us into more subtle levels of experience, so too will the nature of the holding change to accommodate this. We may grow from our first bond with the mother's body through personal, social, and ideological relationships to the experience of being held by universal consciousness.

Whether the student or client is a tiny infant, an older child, adolescent, or adult, a relationship of mutual trust and respect needs to be developed. This requires from the practitioner an attitude of receptivity, humility, and the willingness to "not know" but to listen and learn. At some deep level both infant and adult student may know intuitively whether this particular relationship will serve them at this moment in their life. The practitioner also learns to sense from moment to moment whether the student is ready and willing to actively participate in her own learning, and so comes to trust the student in her own process of development. The possibility for healing to take place occurs within such an atmosphere of mutual trust and openness. Working together in this way, something happens at the moment when minds and hearts meet that I can only describe as unconditional love. An open space is created between student and practitioner and there, with vision, support, and love, the learning and healing can unfold.

This is a small and simple statement, but it expresses the essential heart of Body-Mind Centering and any truly healing work. I have been privileged on many occasions to witness this wordless exchange of love and trust in Bonnie's own work with her students and with the young and often severely damaged children whom she has helped. This love and trust continues to inspire me.

Chapter Six

Deepening Contact
with the Source of Movement

The forces acting upon an organism will determine its structure. As we have seen, movement happens as a response to those forces, both internal and external, to which the organism is subject; in this way both the species and the individual being evolve the forms which best support their survival in a particular environment, and they also give expression to these forces through their structure and behavior. To some degree this form needs to be available to change, to adapt to the continuing newness and variety of the personal and environmental factors that are met with throughout life. No one life situation is ever exactly like another, nor one moment in time identical to the next. In order to free ourselves to act in spontaneous response to the ever-changing flow of life, we need to loosen our concept of our bodily form as being a static and unchanging structure.

At every moment cells in our body are dying, undergoing transformation, and new cells are being created. There is continual movement and reorganization going on at this fundamental level, beyond our conscious knowing, yet within the realm of the wisdom of cellular life. Everything about us, even the structure of our seemingly dense and solid bones, is subject to change and hence to readaptation or transformation. Problems occur when we resist or try to halt these natural processes. If we accept that the process of mind creates the patterns we see expressed through the body, just as the wind creates the changing patterns we see in sand, then we understand that the body will be *re*-created moment to moment in the same familiar forms unless the mind is flexible and can change. The "mind" of each cell, body tissue, and fluid, expressed in feeling states, posture, and movement patterns, is by nature open to the constant flow of momentary change. But if we believe either consciously or unconsciously that our body is solid and unchanging, then our movement and postural patterns will reflect this attitude, and we will truly be stuck with an unadaptive vehicle of expression. More specifically, if the process of mind is reflected in a block to the flow of energy at some level of the body and this has crystallized into a habitual pattern of holding, the movement of the mind within the afflicted body area must change in order to allow the habitual holding to be freed. Contacting and working with the tissues at the cellular level can facilitate such change in mind-body patterns. Once freedom of movement is restored, new choices of response are opened up and we need no longer be stuck within an old habit of reaction that does not allow for spontaneous response to ever changing circumstances.

Our movement and postural patterns can offer explicit descriptions of our psychological process and attitudes of mind at both gross and subtler levels. While we can draw some general correlations based on observation informed by experience (more will be said about this in later chapters), I feel that each individual's own experience of connection between body and psyche is unique; therefore it is generally more helpful and meaningful to allow the student to make personal associations herself. This kind of information and awareness about ourselves can be accessed through developing the receptive, present quality of the "mind" of cellular awareness.

Breathing, as it both functions automatically and can also to a certain degree be brought under voluntary control, can provide us with a bridge between conscious and unconscious processes of body and mind. Paying attention to the process of breathing takes us to this threshold between the conscious and the unconscious, enabling us to perceive what was previously hidden. From this unique standpoint we can consciously choose to revision and recreate that which in our unconsciousness was limiting our aliveness and our freedom to choose and express.

In particular, taking our attention to the subtle movements of internal respiration, the breathing of the cells, can enable us to contact the very origins of both our movement and disease. It is at this cellular level that the continual renewal and reorganization of the body is taking place, so by attending to the life of the cells we are opened to new possibilities. We can recreate the patterns that determine our movement and form, or we can simply allow the patterns to express themselves naturally by releasing whatever has been holding them in constriction. The cells and fluids of the body do not need to hold this trapped energy or the memories and emotions associated with it; nor do they need to be numb, inactive, or unsupportive of the body as a whole. It is only our attachment to the familiar that holds them in this way, bound as we are to our history, our concepts, and our fears.

The principles of this process may sound quite simple, and essentially they are. Yet in practice there can be much complexity, and choices as to how to apply the principles are many and varied. We cannot expect to find one single key, some change that will immediately make everything perfectly free and balanced and well-integrated. There certainly may be essential "keys" specific to each person that can open us to experiences of integration or significantly change our perception of ourselves and our environment. However, in general the process is gradual, one opening leading always toward another, with no final end or goal. We may feel at times the need to rest from in-depth work and integrate what has gone before, or perhaps to create new forms of expression in our lives for the awareness we have awakened.

Perhaps the greatest obstacle to a simple unfolding of potential is the attachment we all carry to what is known and familiar, however painful or unsatisfactory this may be. With that attachment come the fears and the anxieties of the unknown as well as resistance to the changes that might take us there. For most of us such resistance has long since become unconscious habit that may require great awareness and patience to radically alter. As practitioners we may therefore find our work is like peeling off layers, one by one, touching first on those levels at which the student or client is available and prepared to make small changes. The process will gradually and naturally lead us closer to deeper and more fundamental issues if we are patient and receptive to what the student or client is showing us.

We may use the Developmental Movement patterns as a framework or reference point, but we will not necessarily or solely work through them directly. The body is multilayered and movement can initiate from and sequence through the cells and fluids of any and every system and structure, which means that restriction or lack of integration in movement can also happen at any level of tissue. In repatterning movement, therefore, we will be giving specific attention to particular anatomical structures or systems of the body in order to facilitate the ability to initiate and sequence movement freely and allow the full expression of that movement.

When energy is moving in a free and balanced way, all body systems and structures are in a state of communication with all others. This state may be reached through practices such as meditation, yoga, martial arts, singing, or dancing. When a spontaneous free flow of energy and mind-body integration are accessed, there is an awareness that encompasses mind, body, and feeling in totality and transcends all distinctions, from which may emerge an experience of spirit or pure consciousness.

This same experience can occur when we have a cellular level of awareness, not differentiating between body systems but focusing on the essential nature and "mind" of the cell. We come to experience the body as a whole through the living, breathing process of all of its cells, even by focusing our attention on only a particular localized area of cells. Communication and inte-

gration occurs throughout the body through the subtle and responsive movements of cellular activity and through continual
wavelike motions that flow unchanneled through the fluid surrounding all the cells. These very subtle internal movements can
be felt when we return to the state of receptive being; by gently allowing their unimpeded flow we can sense a level of integration and wisdom inherent in the body just as it is, whatever
its apparent condition may seem to be.

As a practitioner, in order to provide the holding needed for
a student or client to settle into this relaxed and receptive state,
we ourselves need to relax into the cellular "mind" within our
own body-mind experience. In doing so we amplify and make
more accessible this quality, which helps the student to recognize
the "mind" of cellular awareness within herself. A sense of safety
is also provided; the student feels herself to be met and held at
the level to which she is going. I recognize that for many people
it is not easy to access this deep state of relaxation and "not-doing,"
especially when alone. The "mind" of the "doing" aspect of the
nervous system may try continually to pull us out of this basic
state of rest and receptivity. This restful place, which is the dark
holding of the maternal matrix, arouses in some people anxiety
and a fear of getting stuck forever in a state of inertia—a fear
ultimately of death. Perhaps our early experiences of the quality of this holding, which we have internalized, will determine
the degree of willingness, pleasure, or fear with which we can
settle into the state of simply being. The role of the practitioner
is to facilitate the student in being here, be present with her as
support, and welcome and guide her gently out into the world
again. In this way the experience can be a rich and rewarding one
that engenders healing of rigidified fears and dissolves old resistances to being with oneself.

In a session with a client we may work primarily at the cellular level. Or we may begin here momentarily, establishing a
connection and listening to what the person needs before engag-

ing in a more active way. This first phase is preintentional: there is no consciously determined
plan; we are just patiently being and listening
through our hands and senses. (Fig. 6.1) When
we begin the process of movement repatterning
we start to differentiate, locating and identifying
specific structures of the body by connecting not
just to the cells in a general sense but to the cells
of a particular organ, muscle, gland, or bone, for

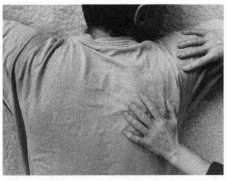

example. Each system and each structure within a system has a
clearly distinguishable quality of energy, density, weight, and movement; we learn to identify each and recognize with increasing
specificity where the movement is flowing clearly and where it
is blocked or too dispersed. The forces acting upon us from both
inner and outer sources do so through all levels of body tissue:
organs, glands, fluids, and nerves, as well as bones, ligaments,
and muscles, each of which either may or may not contribute to
the overall mobility, support, and aliveness of the body. (More will
be said about working with the specific systems in the following
chapters.)

Figure 6.1
Making contact through
cellular touch.

We will be drawn to work in a particular area by what we
observe in a person's movement, posture, and gestures; by qualities she is or is not expressing both physically and verbally; by
information she may give us about feelings or sensations of pain,
discomfort, strength, or weakness, etc., that she is experiencing;
and by the direct information we receive through the contact of
our hands. The way of working will be unique with each person.

We may begin by looking together at anatomical pictures
or models to gain some objective understanding of the location,
function, and relationships of the body parts. This helps to give
the student a visual map of where we will be traveling in the
body. For visually and conceptually oriented adults, this intellectual introduction may feel the safest and easiest. It also stimulates
a sense of active participation in the process to develop. It is not

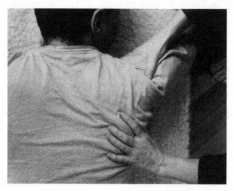

Figure 6.2
Locating and exploring the quality of integration and movement between the bones of the shoulder joint.

always appropriate to begin this way, and of course it is not possible if the student is an infant or young child.

Again we use our hands to define as precisely as possible where in the student's body the structures are located: their shapes, dimensions, and depth within the layers of tissues. This will also give more information as to the quality of integration and freedom of movement of the tissues being contacted. (Fig. 6.2) With the aid of the visual guidelines we have gained from looking at anatomical pictures, the actual structures within one's own body can be experienced. For many of us this first recognition is a moment of revelation—the abstract picture of a bone or organ is transformed into the experience of "my own skeleton" or "my own heart," and suddenly the whole anatomy comes alive. We begin to feel that we are not somehow simply carried around in a body, nor do we carry the body around with us, but we truly and mysteriously live within every cell of our own unique and individual embodiment.

As with working at the cellular level, it is helpful to the clarity of communication if the practitioner now directs her own attention within herself to the body system that she is intending to contact in her student. Her experience of that body system will be communicated through her hands and her presence and so can help the student to identify the experience of that particular system in her own body; we call this resonation. This attention alone can awaken cellular awareness and fuller aliveness in the area. Once contact has been made in this way we can begin to dialogue and to clarify or redefine the flow of movement and the relationships of the parts. This we call movement repatterning.

The Process of Touch and Repatterning

The actual process of movement repatterning involves several stages. In a particular session more emphasis may be given to certain stages than others, depending on the client's needs. However, a general development from a receptive to a more active involvement on the part of the client is usually followed. The practitioner may first work hands-on with the person while in a relaxed, passive, and receptive state. She will use her hands to feed in information about the directions of the new patterning, creating the possibility for openness and integration as needed, through touch, gentle manipulation, and movement of the tissues. This process stimulates new sensations within the body and begins to awaken the nervous system to recognition of the new pattern. (Fig. 6.3) The practitioner must always be mindful to not intrude into depths into which her student has not invited her to go or to force change that is genuinely resisted.

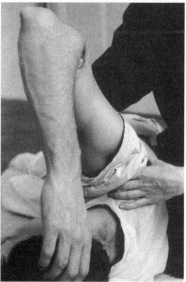

If the student is available to it, the practitioner may give verbal information to guide the movement, such as directions or kinetic imagery relating to the anatomical structures. The student can then join his mind-intent with the practitioner's, actively sensing or imagining the movement while still remaining physically receptive and passive. (Fig. 6.4)

The third stage is to do the movement together, with both student and practitioner active and also receptive to each other. (Fig. 6.5) The practitioner continues with her sensitive guiding, responding to the student's movements and subtly supporting and refining them. The practitioner will offer gradually less and less guidance and support as the student becomes more able to initiate the new pattern of movement for himself.

Figure 6.3
In the first stage of movement repatterning the client remains passive and receptive to the practitioner's touch and directions.

119

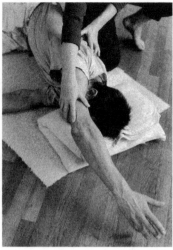

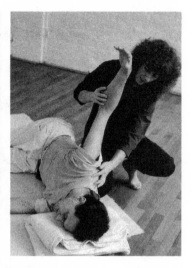

Figure 6.4
In the second stage the client becomes actively involved by focusing on the sensations, images, and directions of the movement.

Figure 6.5
During the third stage the client initiates the movement, with only as much support and guidance from the practitioner as is needed.

Figure 6.6
Less support is given as the client integrates the new patterning himself.

Sensitivity, which comes from using only as much effort as is needed at any moment, is required of the practitioner if she is to ascertain how clearly and actively the student can initiate and carry out the movement to its fullest range without assistance. (Fig. 6.6)

To further strengthen and integrate the pattern, the practitioner may then offer a gentle but firm force against which the student moves. This resistance may come from her own hands or other body parts, from the ground or other surfaces, or from the force of gravity. For example, pushing against the ground as in the Developmental Push patterns helps to integrate the limbs into the spine. (Fig. 6.7) In the final stage the resistant force is released, allowing the movement to be expressed in its fullness of range and energy. With the release of resistance at the appropriate moment, the transformation of a push into a reach can occur. This may be done through integration into specific movement patterns (Fig. 6.8), or through expressive dance or everyday move-

Figure 6.7   The client then moves against a resistant force; this helps to further integrate and strengthen a new movement pattern.

Figure 6.8   Finally the resistance is released, and the range of movement and energy is allowed to express fully in free movement, or integrated into specific movement patterns.

ment activities. Verbal dialogue or counseling may also facilitate integration at this stage.

All structures of the body interact with all others and are part of a dynamically changing whole body pattern. We will therefore find that our dialogue leads continuously from one place to another, one level to another. Flexibility to change and travel where the process guides is required of the practitioner. A change at one level or body area will affect the relationships of the whole. We often find that repatterning through one system will automatically repattern another; for example, change in either organs or ligaments will repattern the use of the muscles. We may also need to follow an unraveling process throughout the body parts, both to free a channel for the expression of the energy that has been awakened or released and to facilitate the alignment and integration of the new patterning into the whole.

We receive information about our inner and outer worlds

and the interplay between the two through three different types of sensory nerves. The "exteroceptors" receive stimuli from the external environment primarily through the skin (regarding touch, pressure, pain, heat, and cold), and through the nerves of the special senses of hearing, vision, taste, smell, and equilibrium. The "interoceptors," located in the internal organs and soft tissues, receive sensory information concerning internal, visceral body processes. And the "proprioceptors," located in joints, tendons, muscles, and ligaments, record information about the position of the body in space, its movement, and the relationships of body parts.

Through consciously bringing our awareness to the specific organs, glands, muscles, joints, etc., we can develop and refine the ability of the sensory receptors of the nervous system to receive and process subtle information about the quality and circumstances of these structures. The sensory functions of the nervous system are developed through conscious use in this way. The nervous system is a primary vehicle through which we perceive and recognize what actually is present and occurring at subtle levels of the body and through which we can consciously redirect or free movement to affect changes in these conditions. This process is called "sensing" in Body-Mind Centering terminology.[1] Sensing has a quality of careful, self-conscious attention that can be both active and receptive, inner- or outer-directed; it is a way of aligning thought and sensation. Through the interaction of the nervous system with the other body systems, movement is repatterned and our general awareness of ourselves deepens.

There is a fine distinction between the subtle directive process of repatterning through the nervous system and the less directed process of allowing a free and spontaneous flow of movement at the cellular and fluid level, which helps release the body into a natural state of aliveness and integration. When the connection between practitioner and student is at its most finely tuned, the distinction between actively guiding and receptively supporting the student's process dissolves. "Being" and "doing" merge into a feeling of connection and integration. In such moments change can be realized; a thought, image, or idea becomes a living and embodied experience. It is a state of mutual awareness and presence of each to herself and to the other, momentarily beyond thought, where will and the intention to act come together in balance with acceptance and surrender to the process and to what is.

It is important in integrating any inner change or new awareness to bring attention back to one's surroundings and simply move in a natural and unselfconscious way, directly experiencing the body in movement. We call this process "feeling."[2] It is an expression of the fluid systems and links us in an active way, through communication and responsiveness, with the world outside. (We will return to this in Chapter Nine.) Although I am primarily describing in this chapter the central process of hands-on movement repatterning, the principles of Body-Mind Centering work have broad application and may also be experienced in personal and creative movement exploration, dance improvisation or technique, and other movement practices or disciplines.

As we dialogue with the physical presence and condition of the body structures and the patterns they are expressing, we may also find ourselves perceiving messages regarding the emotional and psychological states that these patterns are reflecting. We may choose to engage in dialogue through the body with the underlying psychological processes, exploring such questions as "What is this pattern trying to tell me?" "What is it serving?" "How does it feel to be this way?" "What are the attitudes, fears, desires, and needs behind this way of holding and expressing myself?" "Do I accept this condition?" "Am I ready to make a change, and what is needed to do this?"

Frequently, movement patterns will remain blocked until the underlying issues have been addressed in some way and the choices involved made conscious. This can be done through verbal dialogue, imagery, drawing, free movement, and dance expression, as well as through the process of listening, guiding, and responding through the hands. In this way information about emotional and psychological processes can be accessed through the body and also expressed by the body through the creative process of embodiment, repatterning, and integration.

The "Mind" of the Body Systems

Just as each body system expresses itself through a different quality and rhythm of movement, so too does each evoke a particular "mind," a unique and recognizable energy or quality of awareness, perception, and being that is expressed through the individual's presence and activity. These systems and their related "minds" are aspects of ourselves, particular expressions of energy encompassed within our totality. In our search for wholeness we are attempting to embrace and work with each of them.

The movement quality and "mind" of a system reflect its physical structure, functions, and physiology. By directing our attention to a particular system and its attributes of movement, function, structure, and so on, we can initiate movement from there and experience and recognize the "mind" of that system in movement and stillness. An appreciation of the different qualities of experience in our own body-mind continuum will enable us to recognize their expression in others.

When we are in a state of healthy balance, meaning in this context that we can function from an awareness of our totality, expression through all of the systems is available to us. This means that we can respond appropriately to the changing environment, transitioning easily from one quality of "mind" to another, possibly to one in extreme contrast. This responsiveness supports and is a measure of our health. The capacity to be both mindful of the present moment and willing to let go and change in the moment is what allows our energy to flow freely and be continually renewed. Many forms of disease and illness are created by the unwillingness or inability, often at deep and subtle levels, to change, to let go of what was and make space for what is.

In terms of the body systems, change happens from one "mind" to another through the very fine membranous skins surrounding all the cells, structures, and fluid systems. These membranes function not only as boundaries to contain and differentiate but as bridges of communication between one state and another. Indecision and the inability to let go and change our "mind" can mean that we are stuck at the membranes, perhaps conceiving of the boundaries as too solid and impermeable to pass through rather than as bridges of transition and communication. This requires us to make a decision to change; this process of active decision-making takes place at the membranes. (This will be described more fully in Chapter Nine.)

We could say that the process of choice and transition is also a cellular one, with the membrane of each cell acting either as a too-rigid boundary that defends or holds in and withholds (therefore risking inflation or breakdown), or as a vehicle of communication. Perhaps, too, the membranes may give too little containment and definition, creating a sense of unclear boundaries, fragmentation, extreme vulnerability, and sensitivity. Under conditions of stress we may feel ourselves swing between the two extremes. Illness, from cancer at one end of the spectrum to psychotic disorders at the other, may be seen as an expression of the degree to which the body-mind is unable to respond, let go, change, or contain at the membranes of the cells and body systems.[3]

Expression and Support

We each prefer certain systems through which we most frequently or readily express ourselves, and tend to avoid others. Each person will "choose" combinations unique to them; this is usually a natural reflection of personality and not something we should attempt to alter. However, when a particular system or combination of systems is unfamiliar or inaccessible, the freedom to express our totality is restricted. Such systems may be noticeable by their absence in a person's overall expression. We call them the "shadow," and this bears relationship to Jung's conception of the shadow in psychological terms. When we are expressing through a limited area of our potential in this way, greater strain will be put on those systems which are available to us through continual use and demand, and they may in time suffer from exhaustion and stress-related disorders. Bonnie Bainbridge Cohen describes our problems as being our strengths that have not recuperated from overuse and stress.

In order to allow the expressive systems to rest and recuperate, we need to learn to access those systems that are usually unavailable or not fully embodied. This process of bringing the "shadow" systems into conscious awareness and expression may feel difficult or threatening at first. Their unfamiliarity can evoke fear, resistance, and uncertainty. We must approach them with careful attention and only when the student or client is ready and willing. Deep, underlying emotional issues may begin to surface. Some students will choose to use counseling or psychotherapy to support and enrich this depth of work, while for others the caring and attentive presence of the bodywork practitioner may provide the holding environment needed to support and integrate their growing self-awareness.

Accessing all the systems in this way brings a greater range and freedom of movement and expression, and new qualities will emerge. The former shadow qualities can then take on the function of support for those systems through which we usually express, adding depth and richness to them. As we allow the supporting systems to come into awareness, the expressing systems can temporarily take on the supportive role and in this way can rest from their usual expressive activity. We can look from this perspective at recreational activities that tend to revitalize through this rebalancing of systems. There are many forms of recreation and we can begin to base our choices on our individual patterns of expression and support. (This will be made clearer when we discuss the qualities of the individual systems in the following chapters.)

Through awareness and conscious action we can reclaim the dynamic interaction and balance of the body systems and their qualities of "mind" and movement expression that are inherent in each person. Our vision then naturally begins to widen and deepen as we perceive how in our wholeness and individuality we are also dynamically interacting parts of ever more inclusive organizations and communities.

Exploration: Making Contact through Cellular Awareness

The following exploration offers a deepening of the experience of Cellular Breathing, described in Chapter One. As you journey within you may find you can relax and focus more deeply with the support and attentiveness of another's presence. For the facilitator, this process enables you to meet with a student or client in a receptive and unobtrusive way; the quality of simple presence is an essential ground for growth of the healing relationship and is of value in any form of communication.

With a partner you can practice developing the receptive mind of listening through touch. One partner lies down comfortably, allowing her mind to settle into awareness of the process of Cellular Breathing, while the other person sits quietly by her side and focuses in the same way. The latter rests both hands gently on any two areas of her partner's body to which she feels drawn to make contact. Keeping awareness focused lightly on the areas of contact, both partners open their imaginations to the presence of the cells and the pulsing of internal respiration in the areas under and between the hands. Try to let go of any preconceptions or a desire to "do" anything, and simply rest together like this, relaxed but attentive to the presence and activity of the cells.

After a while the hands may want to move to another location, or you may spend the whole time with contact in only one area. And that is all. Just listen and wait, simply observing any sensations, images, and feelings that arise. Observe and let go; don't get caught in following trains of thought or attempting to analyze your experiences. Keep bringing your awareness back to the simplicity of the moment, of cellular presence. You can comfortably continue this for about twenty to thirty minutes. The person lying down may then wish to make some gentle movements to come back into sitting; share your experiences if you wish. The roles can then be reversed.

This is a very simple way both to relax and to make contact with a nurturing quality within yourself. You may also discover some interesting sensations and insights. The important thing here is to try to let go of preconceptions or the desire to act. Simply be as present, receptive, and fully attentive as possible.

The Body Systems

Chapter Seven

The Container:
Form and Structure

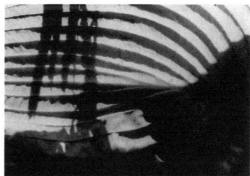

The Skin

When we touch another person's body our first contact is with the skin, the primary boundary that differentiates the physical body within from other physical bodies outside. Through awareness of this boundary, gained through the experiences of contact with the world outside, the infant first begins to identify itself as a unique and relatively separate individual. In this experience are the first intimations of selfhood. As we have already seen, touch plays an essential role even from the very beginnings of intrauterine development. The infant is learning about the world around it as it experiences this world through touch. The development of a healthy ego also depends very much during the earliest phases of life on adequate holding, touch, and the stimulation this gives to the growing sense of body boundary. For adults too, where this experience of self and personal boundary has not been fully nurtured and developed or has been temporarily lost, such stimulation through contact can be beneficial in helping to redefine what is self and what is other.

The skin contains the organs and soft tissues, maintaining their form and retaining body fluids. It also serves to protect the delicate tissues within from injury, harmful bacteria, and excessive radiation from the sun. This last is achieved through the pigmentation that occurs in the outer layer. The outer layer consists of dead cells which are continually being shed and replaced from within, rather like the shedding of the snake's skin. Within the inner of the skin's three layers are deposited fat cells, which provide insulation and shock absorption. As well as insulating against heat and cold, the skin is the organ primarily responsible for the regulation of the body's temperature. Through the skin we also eliminate wastes in the form of sweat. It envelops and contains the whole of us and gives definition to our form. It is the organ of appearance through which we present ourselves to, and are seen by, the world.

Like the membranes within the body and in fact any other natural boundary, the skin also serves as a vehicle of communication. It is an elastic and highly sensitive organ that links our inner world to the outside world and enables us to perceive and learn about both simultaneously. The environment touches our skin, and its qualities are transformed into messages that we interpret as heat, cold, pleasure, pain, comfort, pressure, etc. We can respond to these messages through reflex, instinct, or choice. Through our skin we also touch the world and express something of who we are and how we feel; in this contact we gain feedback from the world about ourselves. As Deane Juhan writes:

> We can never touch just one thing; we always touch two at the same instant, an object and ourselves, and it is in the simultaneous interplay between these two contiguities that the internal sense of self—different from both the collection of body parts and the collections of external objects—is encountered ... my tactile surface is not only the interface between my body and the world, it is the interface between my thought processes and my physical existence as well. By rubbing up against the world, I define myself to myself.[1]

The high degree of sensitivity of the skin is due to the abundance of nerve endings located throughout the skin in its middle layer. The skin and nervous system share their origin in the ectoderm, one of the three primary layers of germinal tissue in the embryo, out of which all the body systems develop.[2] The skin and nervous system are intimately related in their functions of sensitive communication and the transformation of stimuli into perceptions and responses. These perceptions underlie our experience and feeling of self and our relationship to "other."

> This dialectic (between body and the world) is lifelong, and its formative power can hardly be overstated. It establishes preferences and aversions, habits and departures, becomes the very stuff in which attitudes are ingrained. The "feel" in my skin and the "feelings" in my mind, what I "feel" and how I "feel" about it, become so confounded and ambiguous that my internal "feelings" can alter what my skin "feels" just as powerfully as particular sensations can shift my internal states. It is not too much to say that the sensory activity of the skin is a major element in the development of disposition and behavior, an element with enough sophistication and plasticity to account for wide divergences of experience and observation.[3]

The skin differs from the other special sense organs of the nervous system—the eyes, ears, nose, and mouth—in one important aspect. It senses continually its environment, mediating between inner and outer; it does not close. This continuity of sensory perception is necessary to the development of a sense of continuity of self; only with this sense of continuity can a stable and adaptable ego develop. The skin perceives both continuity and the processes of change with which it is continually being confronted moment to moment.

By maintaining awareness of the enveloping skin in its entirety, we can experience inner and outer environments together and the place of the skin as a natural boundary, a membrane that both separates and unites the two environments. A natural boundary, unlike a mentally conceived one, does not divide and restrict but exists as a line of definition and a medium in which potential for change and exchange is boundless. The experience is one of both containment or integration, and expansive spaciousness. It is not dissimilar to the "mind" of cellular awareness but has the heightened consciousness and lightness of sensitivity associated with awareness of the nervous system. Moving with awareness of the skin can give a feeling of lush sensuousness and pleasure.

Exploration

Some specific methods of stimulating awareness through the skin have already been described in the chapters on movement development. To experience the general sense of the skin as the body's containing membrane and as an organ of communication between inner and outer, you can explore as follows:

1. Begin as with the Spinal Push patterns (p. 52), with your forehead on the floor, knees and elbows flexed beneath you. Slowly and gently roll your head on the floor in all directions so that all surfaces of the head and face make contact with the supporting surface. Then allow your body to move freely and change its position as you roll, to let every area of your body surface make contact with the ground. Keep your mind focused on the meeting of your skin with the ground, letting your attention travel as the area of contact changes. Feel that your skin opens and spreads to receive this touch, and try to maintain a quality of relaxed concentration so that your awareness is continuous with the movement. Then rest and hold in your awareness the skin as one continuous and enveloping membrane.

2. With a partner, simply touch their skin with your attention focused at this level. Gentle pinching or sliding of the skin, stroking, and light massage can also help increase awareness. It is helpful to use primarily the fingertips with a light quality of contact. Like the "cellular" contact, this can be a very restful and nurturing experience, but with a particular quality that is light, sensitive, and spacious.

The Skeletal System

The skin is the outermost expression of our bodily form. Let us now look beneath the skin at the densest supporting structure of that form, the skeleton. The skeleton consists of several types of bone, cartilage, and connective tissue. The bones are the most solid of the body's structures, containing deposits of inorganic mineral salts which give them their characteristic strength and durability. Yet the bones are also living tissue. They develop in utero and early life from cartilaginous tissue; calcium and other mineral salts are taken up from the blood by specialized cells and are deposited within the cartilage to transform it into bone tissue. Cartilage is a tough but elastic substance, capable of both bearing weight and a limited degree of flexibility. Examples of it include the discs between the vertebral bodies of the spine and the covering around the articulating ends of most bones of the body.

Bone tissue itself is of different types, according to its shape, location, and the forces with which it must interact. But even the most dense or "compact" bone, such as that found in the shafts of the long bones of the limbs, is living tissue subject to change and renewal. The whole skeleton is in fact entirely renewed over a period of about two years. Its degree of strength, flexibility or brittleness, and to some extent its shape, will also be affected over time by the amount and quality of activity and stress to which it is subject. Blood vessels and nerve fibers run through the bone tissue, providing nourishment for and communication with the cells of the bones. (Fig. 7.1) Through the center of the shaft of each long bone runs a cavity; this space is filled with marrow and lightens the bone.

In the ends of the long bones and the bodies of the spinal vertebrae, where greatest stress is placed in terms of bearing weight, accommodating to sudden changes in position and the pull of muscle attachments, we find a bone formation that provides a combination of great strength, resilience, and lightness. Tiny spikes of bone tissue form a delicate but extremely strong lattice-like design and the spaces between the bone formations are filled with marrow. This is called spongy or "cancellous" bone; being less dense it serves to most effectively absorb the strain and pressure of the forces converging on the skeleton. (Fig. 7.2)

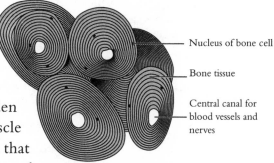

Nucleus of bone cell

Bone tissue

Central canal for blood vessels and nerves

Figure 7.1
Compact (solid) bone.

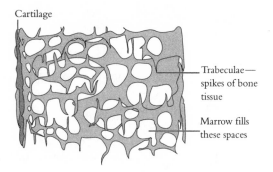

Cartilage

Trabeculae— spikes of bone tissue

Marrow fills these spaces

Figure 7.2
Cancellous (spongy) bone.

The skeleton provides the framework, the inherent structure, of our bodily form, while the fascia, or connective tissue, provides tensile support. (We will look further at the connective tissue in Chapter Nine.) Together, they articulate the balance and alignment of the body's weight with the force of gravity. This allows for both mobility and support, enabling us to rise up from the earth and balance on two feet; this would of course be impossible without a skeleton. The bones of the skeleton also give protection to the soft tissues of the internal organs. The skull cradles the brain and sense organs of the head, the vertebrae protect the spinal cord, and the spine as a whole creates a delicate balance between flexibility, stability, and protection for the nerve fibers that run through it. The ribs wrap around from the spine to the sternum at the front, holding the lungs and heart. The pelvic bones and sacrum together form an open bowl-shaped container for the pelvic organs.

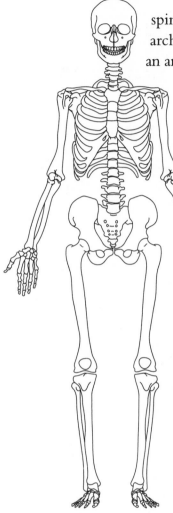

The body weight passes down through the bodies of the spinal vertebrae. The sacrum acts as the keystone of the pelvic arch; we see this reflected in the placement of the keystone in an arched bridge, for example. Through the sacrum weight is distributed between the two halves of the pelvis and down through the long bones of the legs to the feet. Clearly any misalignment of the joints of the skull, spine, pelvis, legs, or feet will affect the alignment and balance of the whole skeleton. (Fig. 7.3)

Yet, as Deane Juhan reminds us in *Job's Body,*

> It is crucial to our understanding of the stability which this unit affords to realize that our skeletons do not support our posture in the same way that flat blocks stacked up on top of each other support a building. There are no flat surfaces or securely stacked members anywhere in our frames. The whole collection of over six hundred bones would simply collapse into a pile if it were not held up by principles very different from those which support a stone wall or a marble column. Bones are not building blocks. They are a complicated and dynamic set of levers and spacers through which the entire musculature can act in order to constantly counterbalance the forces of gravity and of contraction, producing both stable erect posture and freedom of motion.
>
> There is in the skeleton itself nothing inherently upright or even stable. And yet, when it is working in concert with the connective tissues and the muscles, it creates a rigidity without which we could not long survive. . . . Boneless, the whole system would fall in on itself like a tent without poles.[4]

Figure 7.3
The skeleton—front view.

The meeting of the weight of the body with the ground through the feet provides the leverage that propels the body mass through space in walking, running, or jumping. The small bones of the feet and toes all help to articulate this support and transfer of weight in movement, adapting the body's position to the surface of the ground. If the weight is held up out of the feet instead of being released into gravity, and the feet do not articulate in this way with the supporting surface, the strain of maintaining balance in standing and walking will be felt as tension higher in the body in the areas in which the weight is being withheld. Active grounding through the feet develops during the crawling and creeping activity of the infant, where the movements between the bones of the feet lever through into the legs then the spine to push the body forwards. The impulse travels through every joint of the body, from toes to head and fingers. Then, in standing, "If we embrace gravity, the resulting counterthrust upwards from the ground through the head, lifts us automatically."[5]

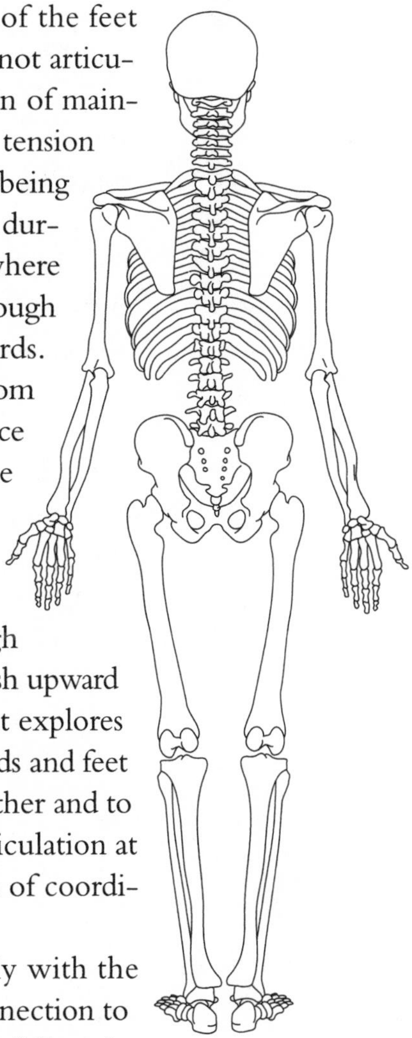

Similarly, the integration of the bones of the upper limbs develops during the early crawling phases as the infant practices releasing weight through the bones of the shoulder girdle, arms, and hands to push upward or backward. The variety of movements that an infant explores in the quadruped posture, where the bones of the hands and feet are learning a multitude of ways to respond to each other and to the supporting surfaces, also helps to develop fine articulation at all the joints. A greater degree of dexterity and range of coordinated movement is therefore possible.

The shoulder girdle does not articulate directly with the spine but has a freer, more spacious, and complex connection to the center than that of the pelvic girdle and legs. If we follow the bony connections from the spine outward, we first travel around the ribs to the front where they articulate with the sternum; the top of the sternum articulates with the two graceful S-shaped curves of the clavicles, or collarbones; these two small sternoclavicular joints are all that connect the shoulder girdle to the rest of the skeleton. (The appendicular skeleton consists of the shoulder girdle, pelvic girdle, and the bones of the limbs. However, in the Body-Mind Centering model, the axial skeleton is

Figure 7.3
The skeleton—back view.

139

comprised only of the most central bones, the skull and vertebrae. The rib cage, like the pelvis, is considered to be a part of the appendicular skeleton, reflecting the importance of articulation of these bones with the spine.) The clavicles act as horizontal struts that permit freer movement of the arms and protect the ribs and thoracic organs from the pressure of the shoulders' weight. The clavicles articulate with the scapulae, or shoulder blades, near to their upper tips so that the scapulae can glide freely over the rib cage. Into the socket at the outer edge of each scapula fits the head of the humerus, the upper arm bone. The movement possible at the shoulder joint, between the two parallel rotating bones of the forearm, among the eight tiny bones of the wrist, and among the long bones of the hand, fingers, and opposable thumb (unique to humankind), gives a great range and complexity of finely articulated movement possibilities. Mobility at all of these joints enables us to express with both precision and feeling, two qualities particularly important to the artist, musician, dancer, bodyworker, and anyone involved in sensitive work with their hands.

The Joints of the Skeleton

"The word 'joint' means to unite. The joint then is the uniting of the two structures together to form a mobile unit."[6] In repatterning through the skeleton we will be looking to see whether the bones at any joint are integrated: do the articulating surfaces of the two bones at any joint have a clear and responsive relationship to one another so that there is a feeling of connection between the two? We will also want to know whether there is a full freedom of movement at the joint or restriction and jamming there. With this we will notice how the bones are habitually aligned and whether this alignment allows for the clear passage of weight through the skeleton. When weight passes through the center of each joint, the range of movement is most free.

There is movement, however slight, at every joint of the body, including the sutures between the skull bones and the joints of the bones of the face. The freedom to express ourselves fully requires that we do not habitually hold at any joint. There is always the possibility of movement and response—a readiness that rides on the subtle movement of breath through the joint, creating a softening and an openness. "The joints of the body are about nothingness ... about the possibilities of movement,"[7] writes Bonnie Bainbridge Cohen.

Within the body there are three main types of joints: the fibrous, cartilaginous, and synovial joints. The flat bones of the skull and the two parallel bones of the forearm and foreleg are connected by fibrous or ligamentous tissue, which allows for limited movement between these bones. The bones of the forearm and foreleg rotate around each other, linked together by an elastic ligamentous sheath. (Fig. 7.4)

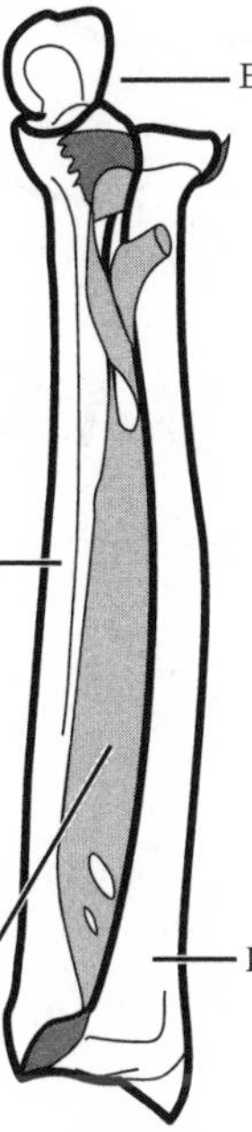

Figure 7.4   A fibrous joint—the forearm.

The bodies of the spinal vertebrae are linked by discs of cartilage that make up about one quarter of the length of the spine and can be compressed or expanded slightly to accommodate movement between the vertebrae in forward and lateral bending (flexion), backbending (extension), and rotation. They cushion the force of weight falling through the spine; but if through misalignment of the vertebrae the weight falls through one side habitually, the disc can become over-compressed there and this may in extreme cases result in a "slipped disc." The whole spine then curves to one side or the other, or to the front or back, compressing on one side and fanning out on the other like an accordion. (Fig. 7.5) If chronic holding patterns set in, the spine may become frozen in a curvature to one direction only, unable to move freely throughout its normal range of flexibility. Freeing movement between the

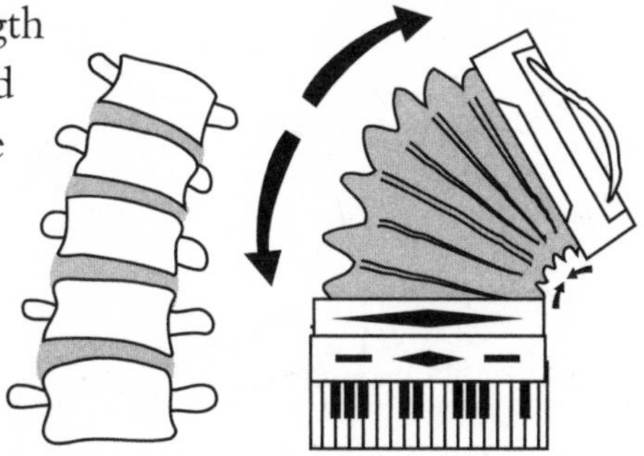

Figure 7.5

The image of an accordion can help free movement between the vertebrae, and balance the force of weight falling through the discs around their center.

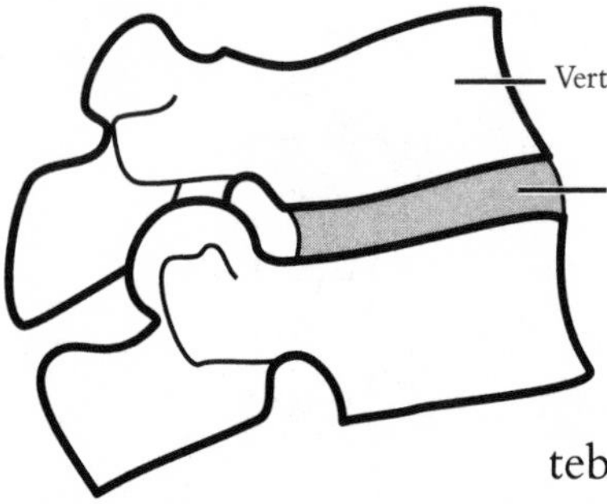

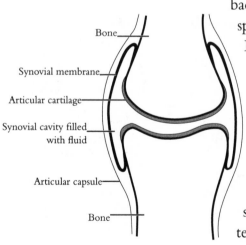

Vertebral body

Intervertebral disc

Figure 7.6
A cartilaginous joint—
the spine.

vertebrae in all directions encourages the weight to fall more centrally through the body of each vertebra. The "body" of the vertebra, the front portion, is the most substantial part of the bone and is designed to carry the weight. Alignment of the bones such that weight falls through the center of the vertebral bodies creates the most efficient means of support and decreases the likelihood of problems caused by unbalanced and excessive compression of the cartilage discs.[8] (Fig. 7.6) We are often unaware of how delicate an art is the balancing of the body in the upright posture. As a species, "we are still in the evolutionary process of evolving towards verticality."[9] In Body-Mind Centering we look for a dynamic balancing of the spine through freeing movement potential, rather than through the imposition of an external image of "correct" alignment.

Cartilage is also found at the pubic symphysis, the joint between the two wings of the pelvis at the front; this allows movement between the two sides of the pelvis here as well as at the back, the sacroiliac joints, where the sacrum of the spine fits into the space between the pelvic bones. Between the ribs and sternum there is also cartilage, increasing the mobility of the rib cage for expansion and contraction during breathing.

The third and most common type of joint is the synovial joint. Here the two articulating ends of the bones are enclosed together within a fibrous capsule. A membrane lining the capsule secretes synovial fluid that fills the joint space; this viscous fluid lubricates the joint, protects it, and acts as shock absorber. The bone ends are covered with cartilage to further protect them from friction. (Fig. 7.7)

A few joints also have within the joint capsule a

Bone

Synovial membrane

Articular cartilage

Synovial cavity filled
with fluid

Articular capsule

Bone

Figure 7.7
A synovial joint—cross
section of finger joint.

142

"floating" disc of cartilage that articulates independently with each bone, creating a double joint action. The articulation of this double action at the joint gives a particular quality of alertness and decisiveness to movements that are developmentally significant, such as in the temporomandibular joint (where jaw meets skull), and in the joint of the wrist end of the ulna (the long bone of the forearm, on the little finger side) with the first small bone of the wrist on that side.[10] (Fig. 7.8)

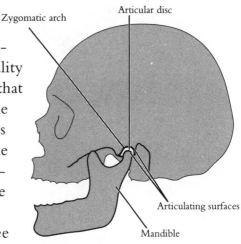

Figure 7.8
The temporomandibular joint (TMJ) showing the articular disc and double action of the joint.

Within the synovial capsule of the knee joint are two semicircular rings of cartilage called the "menisci"; their ends are attached near the center of the upper surface of the tibia (the larger of the two lower leg bones), and their curved middle portions can swing with or in opposition to the direction of movement of the tibia. Freedom of movement in the menisci can prevent their torquing and tearing. When the menisci are freed to swing within the joint space, they contribute greatly both to the knee's ability to accommodate considerable weight-bearing and to its flexibility without damage to its vulnerable structure.

The function of the knee is extremely complex, as this joint takes a great deal of stress in balancing the whole body's weight on a very small area. Traditionally, the knee is considered to be a hinge joint; however, the actual structure can be likened to two very open ball-and-socket joints lying side-by-side. The way in which we conceptualize or visualize a joint affects the quality and freedom of movement there. The idea of a double ball-and-socket joint at the knee can deepen the sensation of an arcing movement, giving more roundness and fullness to the action; it also accommodates the fact that the two sides of the joint (lateral/outer and medial/inner) circumscribe different ranges of movement in normal flexion and extension of the knee. When

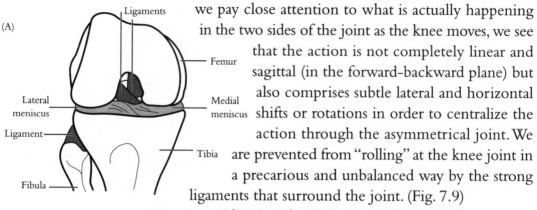

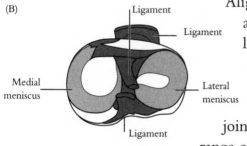

Figure 7.9
The knee joint (A), as seen from the front, and (B), the upper surface of the tibia.

we pay close attention to what is actually happening in the two sides of the joint as the knee moves, we see that the action is not completely linear and sagittal (in the forward-backward plane) but also comprises subtle lateral and horizontal shifts or rotations in order to centralize the action through the asymmetrical joint. We are prevented from "rolling" at the knee joint in a precarious and unbalanced way by the strong ligaments that surround the joint. (Fig. 7.9)

Aligning the skeleton in the vertical posture is an art of fine balancing which is aided by the ligaments. Ligaments are made of very strong but flexible connective tissue; they join bone to bone over the joint spaces, strengthening the joint capsule. They give stability to the joints but are elastic enough to permit the fullest range of movement possible without compromising balance and skeletal integration. The ligaments align the bones in a dynamic way, giving direction to the bones' movement and tensile support for the maintenance of posture.

Whenever we work with the skeleton, we are in fact repatterning through the ligaments. They create a complex web of threads throughout the skeletal structure, supporting at every location and angle where there is stress from the forces of movement or weight falling into gravity. They function rather like the guy ropes of a tent, pulling on all sides of the bones at a joint to maintain the natural and central position of the bones. Like muscles, the ligaments can be under- or overused, too highly or too poorly toned, tightening to pull the bones out of their natural position where they have become shortened and hardened or not giving enough support where they are flaccid and so allowing a collapse through the bones at that place. In some traditional approaches, practitioners have been taught that overstretched and

lax ligaments cannot regain their tone. In Body-Mind Centering we find that with awareness and reeducation through touch,
the ligaments can regain appropriate tone; they can then more
fully support or release the bones back into their natural alignment. As with other tissues, we pay attention to the "mind" of
the ligaments in order to effect change.

Principles of Working with the Joints

For movement to have the full and circular quality that is its nature,
an equal space between the two bones within a joint needs to
be maintained throughout the action: "[T]he balancing of the different joints is about maintaining an even space in the joint
throughout the full range of motion of the bones involved. By
keeping one's awareness on the space, jamming can be eliminated and movement continues into external space."[11]

    We can facilitate this opening of the space by first
taking out the slack between the two bones, imagining
drawing them apart like stretching two ends of a piece
of elastic away from each other. We may then apply the
image of two spheres, or cogwheels, that correspond
to the articulating ends of the two moving bones, rotating counter to each other. This counterrotation of the
bones is what happens naturally in movement, but by
using this image to actively and consciously direct
the movement, the rotation can be freed into its fullest
range and the space within the joint maintained to
prevent jamming. If jamming within the joint occurs at
some point, then the resulting form of the movement will
be angular and inhibited, with the moving bones pulled in,
instead of being allowed to circle or glide freely. (The friction
caused by continual jamming may eventually wear away the protective cartilage at the ends of the bones; this is one causative factor in arthritic conditions.) (Fig. 7.10)

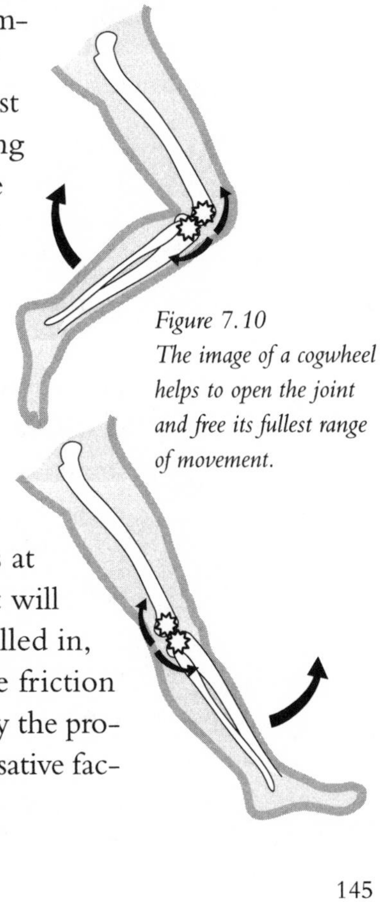

Figure 7.10
The image of a cogwheel
helps to open the joint
and free its fullest range
of movement.

145

We might further refine the action at a specific joint by moving only one of the two bones involved. So that the nonmoving bone is not carried passively with the moving bone, resulting in a lessened clarity and articulation of the joint, we first apply the direction of a countermovement in the stationary bone. This is not an actual movement, but the feeling and intention of a force moving in a direction exactly opposite that of the moving bone. This force anchors the stationary bone so that it acts as a stable support for the moving bone. Support precedes movement; it is important to establish the feeling of counterthrust in the nonmoving bone before beginning the actual movement.

The principle of support preceding movement is one we use widely, and it also applies to the emerging of the Developmental Patterns. The infant first finds its place at the new level, then begins to move out from it. This means that we first know our ground and our place in space in order to have a supportive base from which to move out from and back to—a central alignment with the earth. Such grounding enables movement to be expressed with a clarity of articulation in space. Without grounding and a clear spatial sense we easily become lost as we move. The skeleton when well integrated expresses this clarity of spatial form in movement.

The technique of feeding in a gentle and rhythmical compressive force or weight between the two bones of a joint (as described earlier in reference to the Push patterns) helps to integrate the skeleton and give a sense of connectedness to the ground. The compressive force awakens "recognition" between the two bones involved through stimulating the proprioceptive nerves in the joint and its surrounding ligaments. It also helps to balance the muscle tone around the joint. The resulting feeling of groundedness gives a stronger sense of personal and body boundaries. If this pressure is applied with a loose and jiggling action, it can also stimulate movement of the synovial fluid within the joint capsule, which helps to release qualities of dryness and rigidity in the joint movement.

Proximal and Distal Initiation

Any movement at a specific joint in which one bone moves and the other acts as a nonmoving support can be described as initiating either proximally or distally. The proximal end of a bone is closer to the center of the body; the distal end is further from the center and closer to the body's periphery. In defining a movement's initiation in this way, we are looking at the end of the moving bone that articulates within the joint. If it is the proximal end of a bone that moves at the joint, then we call this a proximal initiation of movement at this particular joint; if the distal end of one of the two bones is where the movement is happening, then it is called a distal initiation.

For example, hold your upper arm still and move the forearm up and down from the elbow joint; the proximal end of the bone of the forearm, the ulna, is circling around the distal end of the nonmoving bone of the upper arm. This is a proximal initiation of movement of the forearm, at the elbow joint. If we reverse the action, still flexing and extending at the elbow joint but placing the forearm on the floor so that it acts as the nonmoving support, the movement will initiate at the distal end of the upper arm bone; this is a distal initiation at the elbow joint.[12] (Fig. 7.11)

In the same way we can rotate the thigh bone in the hip

(A) Proximal initiation

(B) Distal initiation

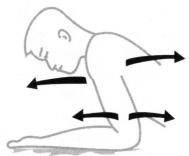

Action happens at proximal end of the moving bone

Action happens at distal end of the moving bone

Figure 7.11
Initiation of movement at the elbow joint.

socket, or excursion the hip socket and pelvic bone around the stationary thigh bone. The first will be a proximal initiation of movement, and the second a distal initiation, both at the hip joint. The dancer working at the ballet barre will utilize both of these actions if her movements are articulating clearly at the hip joint. However, if the action at this joint is not clear, full rotation or excursion may be missing, causing the movement to be initiated elsewhere in the body such as in the knee or lower back. This can create strain and torquing in these areas.

During its development, the infant first learns to initiate controlled and voluntary movement at each joint distally, with the limbs as support against the floor or other surfaces as the center moves. As muscular strength, gross coordination, and extension of the limbs develop through the practice of crawling and creeping movements, more finely articulated movement initiated proximally is being mastered. We see this primarily in the many detailed activities performed through the hands, and in the responsive movements of the head as it is guided by the senses. In dance or gymnastics, for example, the potential for articulated and expressive movements of the feet and legs are also developed more fully.

Movement at any joint can be initiated both proximally and distally and a fullness in the quality of movement can be attained through the balance and integration of these two modes. As adults we may tend to initiate movements proximally with the arms and legs moving out of the central support of the spine, more frequently than distally as in the infant's crawling patterns. If a particular joint is habitually used in one way only, we find that reversing the pattern of initiation can dramatically release tension and open a fuller range of movement; in doing this, it is as if we bypass the habitual pulls and resistances, the unconscious grooves which our movement has fallen into, and we enliven the awareness of the whole area. The habits of mind that underlie the patterns of initiation will also begin to loosen their hold and change as a new way of initiating is experienced, or rather, as

an old way is remembered. Returning to the usual way of initiating, we will find that the movement now has more freedom
and support. Bonnie Bainbridge Cohen writes of one aspect of
the "mind" of these two patterns: "The proximal initiation reflects
in a mind state of gathering in to oneself what is without. The
mind state which accompanies distal initiation is that of being
drawn into space."[13]

Above are examples of some of the basic Body-Mind Centering principles that are used in repatterning through the skeletal system. A more comprehensive presentation is beyond the
scope of this book, but forthcoming manuals from the School for
Body-Mind Centering will give a fuller account of these principles as they are applied in working with the various body
systems. (See page 333.)

The Floating Bones

There are also several "floating" bones that are not attached
through joint connections to the skeleton as a whole but still
articulate with it. These bones serve important functions in relation to balance and mobility.

The patella, or knee cap, is located within the tendon of the
muscles of the front thigh which runs over the front of the knee;
it acts as a pulley for these strong muscles and protects the delicate knee joint. It is attached to the tibia through this tendon,
and articulates with the femur, or thigh bone, in flexion and extension of the knee.

The hyoid bone is a small horseshoe-shaped bone at the
front of the throat above the vocal organs and is attached to the
spine, skull, sternum, and scapulae by muscles and ligaments that
radiate from it. The root of the tongue is attached to the hyoid,
and so it acts as a base of support for the tongue, for the development of early mouthing movements (sucking and feeding),
and for speech. The hyoid bone can be articulated against the

spine and other bones of the skeleton as if there were an actual joint, and its correct alignment gives considerable support to the balance of the head and spine.

Within the ear are three tiny bones that play a part in the transformation of vibration into what we hear as sound. These bones articulate with each other and we can refine our alignment by articulating them with the skull and spine using the same principles outlined above.

The Layers of Bone

In working with the principles outlined above we are generally focusing on the bones in their entirety, in relation to one another and to gravity. We can also take our attention to the layers within a single bone and work with these directly; there are three, and each has its own distinct "mind." When we sense through the skin, connective tissues, and muscles to bring our focus to the bones, the first layer we contact is the periosteum, or "skin" of the bone. This tough, fibrous sheath covers the entire bone and is continuous with the connective tissue sheaths that unite every part of the body. The periosteum also penetrates within the bones to line their hollow cavities; here it is called the endosteum.[14] The ligaments and tendons attach to the periosteum rather than directly to the solid bone beneath, providing a very secure attachment for muscles and support for the joints. Through the continuous connection of ligaments, periosteum, and endosteum the whole skeletal structure is held together and integrated.

The periosteum is filled with nerves and blood vessels; it provides nutrients for the bone tissue and also is important in the process of bone growth and repair. Along with cartilage in early life, it establishes the outer form of the bone and shapes its growth. A tear in the periosteum, therefore, will result in irregular growth of the bone as a whole.

Skeletal stress and misalignment affects the connective tissue sheaths around the bones, and when we work with joints and ligaments we may also extend our focus through the periosteum. In this way we can facilitate the release of tension, torquing, and pulling at that level or enliven the tissues so that they can better support the integration of one joint with the next through the length of the bones. Focusing at this level can also help to differentiate muscle from bone and so increase articulation between them. As the periosteum is affected when the bone is injured or broken, the healing of injuries can be supported by attention to the flow of energy through the periosteum as well as the bone tissue itself.

Healthy periosteum can often be felt to slide over the solid bone that it covers. In contacting the periosteum, however, we might find that it feels glued to the layer beneath, lacking a clear identity of its own. Or, at the other extreme, we may experience this layer as either overdefensive or oversensitive. Some people may find it a relief to drop beneath this highly innervated tissue into the deeper layers for recuperation.

The aspect of bone with which we tend to be most familiar is the actual bony tissue, the most solid layer. Although it is mainly mineral in composition, this is not the bleached, dry bone we may imagine from looking at skeletons. This layer is approximately 25 percent water, and the many small arteries coursing through it give it a pink hue.

Bone is associated with deep internal support. It is the oldest of our tissues, our oldest imprint, composed as it is of the minerals of the earth. For this reason there might be an experience of connection with ancient ancestry when one brings awareness to this layer. We can sense through touch tremendous variation in the density of bone in different people—and sometimes within one body—from soft to brittle; this may change surprisingly quickly as we bring awareness to this layer and balance it with the other aspects of bone. This is an important tissue to work with when there have been fractures; the bone as a whole will often lengthen and the parts reposition themselves.

In supporting the healing of bone injuries, stress, or trauma, we can also focus on the bone marrow. The marrow lies within the hollow cavities of the long bones and also fills the spaces in porous cancellous bone; this is found in the ends of the long bones and in the vertebral bodies, for example. In infants and children red marrow predominates; as we grow older much of this is replaced, especially in the shafts of the long bones, by fatty yellow marrow. The cancellous bone, however, remains filled with red marrow. We might think of the marrow as the molten core or the river within the bone. And while the periosteum and related tissues connect one bone with another through the joints, the marrow is our means of energetic connection between the bones. The art of *t'ai chi ch'uan* works directly with the flow of energy— *ch'i*—through all three layers of bone. Through practice it is said that *ching ch'i,* the essence of life, is mobilized, causing the tendons and ligaments to conduct heat through the periosteum sheaths and into the bones. This changes the constitution of the marrow. Over a long period of time this repeated process causes the bones to become "indestructible, tough, and resilient, not brittle or weak, but as supple as an infant's."[15]

To contact the marrow we move our attention from the periosteum through the solid bone tissues to the spaces within. In the shafts of the long bones there is a thin layer of red marrow lining the inner cavity of the bone; this can be felt as a transition place into the yellow marrow, which lies deeper, or directly into the blood which the marrow constantly supplies with new cells.[16] Strong currents of movement, a sense of "streaming," can often be felt within the marrow. When we support this with our awareness the marrow can initiate an unwinding process that takes with it the other layers of bone. The bone as a whole may be felt to bend, spiral, or lengthen. This "bone-bending" may be initiated, however, at any of three layers; as we simply pay attention, one of them generally initiates the unwinding process. Then ensues a dance in which we alternately guide and follow, allowing the tissues to move toward greater balance, releasing or channeling energy flows that have been disturbed by injury, stress, or strain within the bone.

The marrow may be experienced anywhere along a continuum, from weak and lacking vitality to turbulent and uncontained; in the latter case more awareness of the compact bone is usually helpful. Osteoarthritis and osteoporosis are conditions in which moving attention into the marrow is usually beneficial and can encourage a more fluid and resilient quality in the bones.

In the marrow deep dreamlike states are sometimes experienced; these can be immeasurably relaxing, even blissful. Here everything seems to flow without structure or boundary. Some people feel that energy begins to drain away if they stay there too long. In this case, returning through the red marrow layer to the blood or to the solidity and structure of the compact bone is usually helpful; from here we may move our awareness out to the skin again, through the intermediary tissues. Alternatively the flows of energy experienced in the marrow might be embodied in movement, bringing in other body systems such as muscles and blood flow, to further release any stress, torquing, or blockage within the layers of bone. In general, it is important after contacting any bone layer to integrate that with the others.

Ultimately we are seeking the alignment, integration, and aliveness of all three layers of bone.

"The feeling of bone that is optimally balanced is characterized by clarity of form in the periosteum, resiliency and strength in the compact bone, and focused vitality in the marrow."[17]

Exploration: The Skeletal System

To awaken the living experience of the presence of our own skeletons, we begin by locating and tracing the bones and joint spaces; we use anatomical pictures or models to help identify and clarify their locations and shapes. We then open to imagining, sensing, and feeling their actual presence and movement. Awakening our awareness to their presence within us gives a sense of inner structure and support. Attention directed to the tissues of the bones and the joint spaces stimulates their own "cellular intelligence" so that they may "know themselves." Through our awareness they are energetically activated into providing a deeper and clearer sense of support, internal integration, and grounding.

Exercise 1. Sensing the Bones

To work directly with the skeleton we need first to know how to make contact with it and recognize its qualities. Begin by simply touching your own or a partner's forearm lightly with the tips of your fingers and thumb. Feel the skin, then let your focus and your touch sink deeper, through the layers of soft tissue between the skin and bone, until you have a sense of the shafts of the long bones of the forearm running right through the arm's center. Your touch will become firmer and more precise, but do not press or hold tightly—it is through your mind that you are making contact. Try to sense through the bones of your own body, maintaining the light sensitivity of the nervous system rather than using muscular force to press through the tissues. If you are patient you will feel your partner opening to experience this level herself and allowing you to meet her there.

Maintain this level of contact for a while and be aware of what you are experiencing in terms of bodily sensations, feelings, quality of perception, and attention. Practice sensing the bones in different areas of the body; use anatomical pictures of the skeleton to guide you if it is not already familiar in order to become acquainted with the quality of touch and "mind" of the skeleton.

Exercise 2. Moving Through the Bones

Keeping your mind focused on the bones and their movement, explore initiating movements proximally and distally at each joint of the body. First one bone supports while the other moves, then these roles are reversed to balance action in the joints. The idea of a countermovement in the supporting bone, opposite to the direction of movement in the moving bone, can be used here to give more clarity. (Fig. 7.12)

A.   The image of the cogwheel can be applied at any joint. Try flexing and extending your knee; then, at the articulating ends of the two bones (the femur and tibia), visualize and try to feel the action of two counterrotating wheels that "carry" the circling action of the bones. Maintain this image throughout the whole range of movement and a little beyond. (See Fig. 7.10)

B.   In different positions, from lying to standing sense the falling of weight through the bones. Find the vertical axis through which the weight passes down the center of each bone and joint, as you change your position and relationship to gravity.

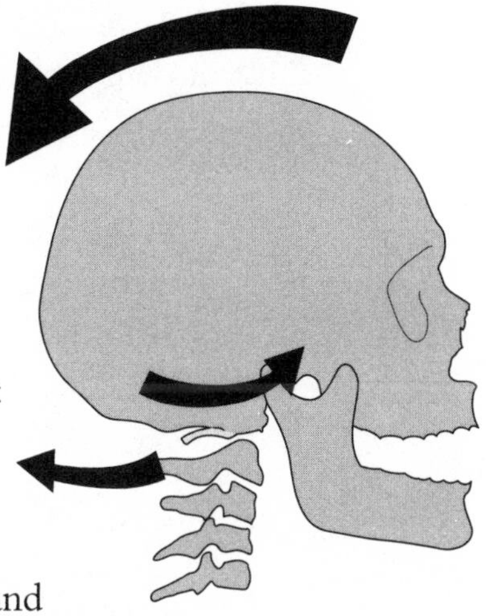

C.   With your mind still focused on the bones and holding in your awareness the skeleton as a whole, initiate movement from them, improvising freely. Move through the bones to discover the qualities that they express. Include in your awareness and movement initiation the ligaments that connect over and around the joints, then the synovial fluid within the joint capsules; notice how each of these affects the movement quality and "mind" of the skeleton.

Figure 7.12
An imagined counter-movement in the first cervical vertebra gives clearer articulation and support to the movement of the skull.

The "Mind" of the Skeletal System

Your experience of the "mind" and movement qualities of the bones, ligaments, and synovial fluid, and the feelings or attitudes you have toward these experiences, may differ from those of others according to the relationship you have to your own skeletal system. In general, however, we find that a sense of form, clarity in space, and light, effortless but grounded movement is expressed through the bones. The ligaments refine this spatial clarity, adding a sinuous, tensile quality of direction and connectedness throughout the skeleton that extends our movement out into space. The "mind" is one of attentive, refined clarity, and precision. The synovial fluid of the joints has a juicy, rebounding, jiggling, throwaway quality to its movement, which is unformed and arrhythmic; it connects us to laughter and a carefree attitude. The synovial fluid balances the relative rigidity and structured quality of the bones, and the clearly defined articulation of the ligaments. Awareness within the skeletal system, which provides us with inner structure, can give a feeling of security, clarity, grounding, and spatial form.

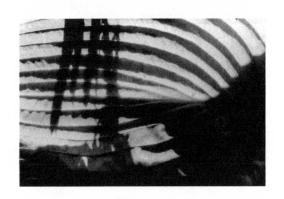

The Muscular System

Many organs of the body are muscular, such as the heart, stomach and uterus, and within the circulatory system the walls of the blood vessels are also composed of muscle tissue. These muscles are not normally under voluntary control; their activities are regulated automatically by nervous and chemical stimuli, in response to inner and outer environmental changes. They are called either "smooth" or "cardiac" (heart) muscle.

When we speak of the muscular system we are referring to the skeletal or "striated" muscles. These attach to and move the bones, crossing over one or more joints, and they are the primary movers of the body. Their work concerns activity, effort, endurance, strength, and vitality. The body's musculature creates the visible shape, or form, of the body and provides a certain amount of support and protection for the soft tissues within. It is a complex system consisting of over 700 individual muscles. In *Job's Body,* Deane Juhan describes the muscular system as functioning like one single muscle with many compartmentalized parts.[18] This concept reflects our experience in applying Body-Mind Centering principles. We find that focused work on specific muscles can affect the musculature throughout the whole body.

An immense amount of energy is required simply to keep the body balanced upright, as the muscles are continuously making tiny contractions to maintain the alignment of the skeleton. Even in the deepest stillness there is a level of minute muscular activity, so that posture is never static but rather an act of subtly dynamic balance. Individual muscle fibers alternately contract and rest, so that the muscle as a whole can maintain its activity. An equal contraction of the muscles around a joint, where balance is maintained but no movement takes place, is called an isometric contraction. If you stand very still and relaxed for long enough you may be able to feel this "small dance" within the stillness.[19] Going through this stillness to the activity at its heart can release you into a free flow of spontaneous movement.

In movement an even greater amount of energy is used, and this also produces a large proportion of the body's heat. The muscles are supplied with an extensive network of blood vessels which provide the necessary oxygen and nutrients for the production of energy in the cells; they also take away waste substances and circulate the heat produced throughout the body. A network of nerve fibers also runs throughout each muscle, extending to and from the spinal cord and brain. Impulses from the "motor nerves" stimulate activity in the muscle, and the "sensory nerves" convey messages back to the brain regarding the activity and state of the muscle. This two-way system of communication facilitates finely tuned voluntary control of the muscle action. The skeletal muscles are also known as "striated" muscles because of their striped appearance, or "voluntary" muscles because their activity is initiated through conscious and willed intention.[20]

A skeletal muscle consists of many elongated muscle cells, or fibers, that are sheathed together in parallel bundles; there will be a number of these bundles in a muscle. They run lengthwise along the muscle, which is also the direction of contraction and lengthening. Each cell, bundle of cells, and whole muscle is wrapped in a sheath of connective tissue. At each end

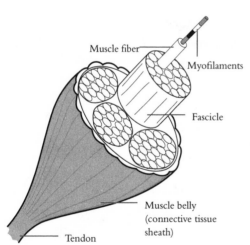

Muscle fiber

Myofilaments

Fascicle

Muscle belly
(connective tissue
sheath)

Tendon

Figure 7.13
The structure of a
skeletal muscle.

of the muscle these sheaths join and elongate into a very strong and resilient tendon that attaches into both the bone and the bone's own connective tissue sheath, the periosteum. (Fig. 7.13)

Eccentric and Concentric Contractions

Within each muscle fiber are many microscopic protein filaments, of two kinds, which lie together in small units, also parallel to the length of the muscle. When the muscle length decreases in contraction, one group of these microfilaments slide toward each other, slipping between those of the other group, so that the individual units of filaments, and hence the muscle as a whole, become shorter and thicker. In lengthening, the filaments slide apart again, decreasing the width of the muscle, muscle fibers, and units within the fibers. (Fig. 7.14) It is similar to the action of placing your hands in the same plane, with fingertips touching, and sliding the fingers of one hand between those of the other so that they lie parallel, then sliding them out again until the two hands separate.

In Body-Mind Centering work we consider that both the shortening and lengthening phases can be active contractions; they are called, respectively, concentric and eccentric contractions, or actions. In any movement, the muscles lying on one side of the joint may be actively shortening or concentrically contracting as those on the opposite side of the joint are actively lengthening or eccentrically contracting; these actions happen more or less simultaneously as the movement is made. The terms "concentric" and "eccentric" contraction are found in traditional terminology, for example in Thompson's *Manual of Structural*

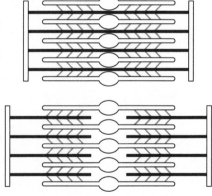

Figure 7.14 Movement of myofilaments during the lengthening and shortening of a muscle.

Kinesiology,[21] but do not seem to be widely used. The specific way these terms are understood in Body-Mind Centering theory has evolved out of two concerns: first, that we study the relationship of mind to patterns of movement initiation; and second, that the force of gravity is taken into account as an essential factor in defining muscle function. Both of these factors will be clarified as we proceed.

Traditionally, the lengthening phase is considered to be a relaxation of the contraction, which allows the muscles on the other side of the joint to shorten. To understand why we might consider the lengthening to also be an active phase and not merely a releasing of contraction in the muscle, try holding your arm up level with the shoulder; then relax all the muscles of the arm and shoulder completely. Because of the weight of the arm and the force of gravity it will simply fall or collapse in an uncontrolled way. If you pull it down by concentrically contracting the muscles underneath the shoulder joint, the arm will be brought down with too much force and speed. (One of the problems that manifests in some spastic conditions is that the modulating of movement through eccentric contraction is lacking.) In order to lower the arm with control, the muscles that course over the top of the shoulder joint in particular must remain actively involved throughout the whole movement. You could think that this is simply a gradual letting go of the contraction; in this case there would still be an active involvement of the concentric contraction which is being gradually reduced throughout the movement, and in fact that may be how we often lengthen our muscles. In doing this we are performing the movement with a "mind" of concentric contraction, which creates a tendency to pull in towards the center of the body, to withhold our energy. A tense, contracted condition can develop in the muscles if the "mind" of concentric contraction habitually dominates movement, because the muscles are not being allowed to experience their fullest possible length in movement and rest.

If we focus on the set of muscles that are lengthening in a particular movement, instead of those that will be shortening on the opposite side of the joint, and actively create the sensation of extending through the whole length of that muscle, we can in fact increase the contractility and the natural resting length of the muscle and free it from a state of habitual contraction or tension. The more specifically we are able to focus on the sliding action of the protein filaments within the muscle fibers, the greater the change in the muscle will be. In doing this we are changing the "mind" of the muscle; the "mind" may feel more open, expansive, and light, or closed, compact, and integrated, according to how we habitually use our muscles and where our awareness, consciously or unconsciously, is focused.

Of course, if a person tends to feel too expansive or unintegrated, or lacks good muscle tone and coordination, concentration on the concentric phase of contraction may be more appropriate and helpful. Another person whose movement tends to be tense and bound might benefit from focusing on the eccentric contraction and state of mind. Ideally we want to create a balance between the two types of contraction so that the range and quality of movement is full and no one set of muscles is overworked to the exclusion of another. Wherever there is overuse there will usually be a complementary set of muscles on the other side of the joint, or joints, which are under-active and require conscious direction to awaken them. The eccentric activity of the muscles extends us out into space and creates spaciousness within; the concentric contraction gives the feeling of drawing in to the center and integrating. (Fig. 7.15)

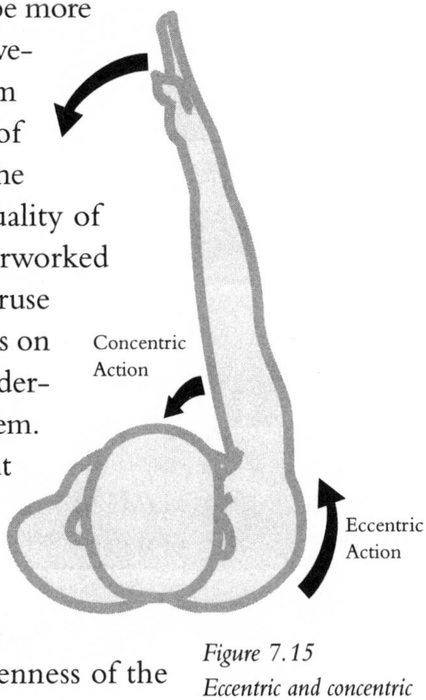

Habitual holding patterns in the muscles obviously restrict the range of movement and the openness of the joints. Energy is being used to hold a muscle in constant contraction, the elasticity of the muscle is lost, and there is less energy

Concentric
Action

Eccentric
Action

Figure 7.15
Eccentric and concentric contraction in adduction of the arm.

161

available for activity so that we become more easily tired. It is good to stretch the muscles to their limit, and even gently extend this limit, but if we attempt to release such a holding pattern by forcefully stretching and pulling the muscle beyond its limit, we may succeed in doing so by tearing the tiny muscle fibers. They will soon form again, but stretching in this way does not fundamentally change the muscle length and requires daily practice, and tearing, to keep the apparent length. If we don't change the way the muscle is being used, its "mind," we will always be battling against ourselves. Our conscious mind may be directing the muscle to lengthen, but the unconscious neuromuscular pattern that actually determines the performance of the movement is holding in and contracting. Moshe Feldenkrais writes:

> It is often enough for a man who is doing something to simply ask himself what he is doing in order for him to become confused and unable to continue. In such a case he has suddenly realized that the performance of the action does not really correspond to what he thought he was doing. Without awakened awareness we perform what the older brain systems do in their own way, even though the intention to act came from the higher third system (of the brain). Moreover, the action often enough proves to be the exact opposite of the original intention.[22]

If we can bring awareness to the unconscious patterning and introduce a new message, a conscious repatterning at this level, we can coordinate the neuromuscular programming with the intended movement. Through this active focusing of attention and awareness, energy and intent can be brought into alignment in the performance of the action.

The Diaphragms

The principles of eccentric and concentric contractions can be applied to any muscle or group of muscles within the voluntary muscular system. This exploration will be further refined in the next section. But first let us look at the specific and perhaps less obvious, though significant, application of eccentric and concentric contractions to the muscular diaphragms of the body.

The respiratory diaphragm is the essential muscular and tendinous organ whose contractions stimulate inhalation and exhalation in the process of breathing. It is, strictly speaking, an organ of respiration, and its contractions happen automatically without the need for conscious intention or direction. However, its actions can also be voluntarily controlled to some degree; it is therefore a bridge between conscious and unconscious processes and an important link between body and mind. Like any muscle of the body, its full functioning depends on its degree of elasticity and the balanced alternation of eccentric and concentric contractions. Restriction in its movements will affect the full and free flow of the breath.

The diaphragm spans the space between the front, back, and sides of the lower rib cage, separating the organs of the upper and lower body like a double-domed umbrella. The outer edges of the diaphragm are muscular and attach into the inside of the lower ribs and front of the spine. Its central part is a tendinous sheath, through which the main blood vessels, esophagus, and nerves pass. On the inhalation the muscle contracts and pulls the central tendon downward, also pushing the lower rib cage outwards; this lengthening and widening increases the space within the upper chest, and the lungs expand, taking in air. This is a concentric contraction, as the muscle is shortening. On the exhalation it eccentrically contracts, lengthening to release the central tendon upwards into the rib cage area. The ribs also fall back towards the

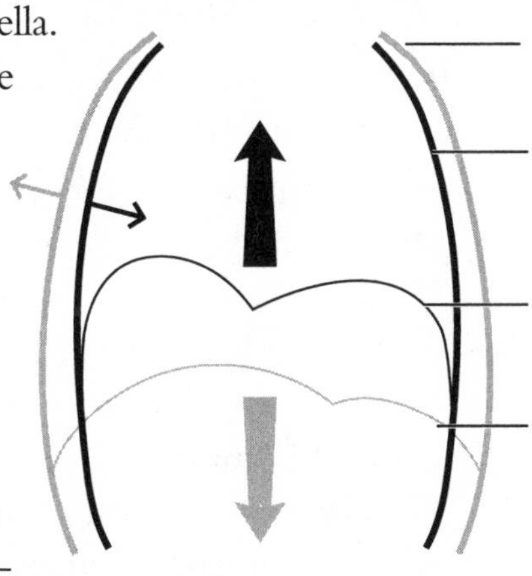

Position of rib cage at the end of inhalation

Position of rib cage at the end of exhalation

Diaphragm in eccentric contraction—exhalation

Diaphragm in concentric contraction—inhalation

Figure 7.16
The thoracic diaphragm—eccentric and concentric contractions during respiration.

163

center to narrow the rib cage. This decreases the space within the lungs and air is expelled from them. (Fig. 7.16) By focusing the mind on the concentric and eccentric contractions of the diaphragm, the action of breathing can become fuller, deeper, and calmer. Diaphragmatic action is also supported by the activity of the intercostal muscles and muscles of the abdomen.

There is another muscular diaphragm in the body; this is made up of the pelvic floor muscles, which lie parallel to the thoraco-abdominal diaphragm, also more or less in the horizontal plane, and can be felt to be an underlying support for it. They connect the two "sitting bones" at the base of the pelvis with the coccygeal (tail) bones of the spine and the pubic bone at the front of the pelvis, thus forming a diamond-shaped "floor" at the very base of the torso. When well toned, this muscular floor is felt to dome upwards slightly, and it gives an important support to the organs lying within the pelvic cavity. When the action of the respiratory diaphragm is free, it gently massages the abdominal and pelvic organs through its rhythmical compression and release; this can be felt all the way down into the pelvic diaphragm, which supports the breathing process by a simultaneous alternation of subtle eccentric and concentric contractions.

Again, balancing eccentric and concentric contractions of this muscle helps to tone and strengthen its supportive function. This can be done by focusing on moving the bones to which it is attached towards and away from each other or by actively contracting the pelvic floor muscles while visualizing or feeling the upward doming increase and decrease with the concentric and eccentric phases of the movement. This is a particularly helpful practice for women in pregnancy and also after giving birth, when the pelvic floor muscles have been stretched so much that they may not easily regain their natural elasticity and supportive function and may tend to collapse downward.

Above the respiratory diaphragm at the level of the uppermost ribs and sternum is another "diaphragm" known as both the "thoracic inlet" and the "thoracic outlet." It is primarily composed of fascia, or connective tissue. Freedom of movement through this diaphragm is essential to the unrestricted flow of blood, lymph, and cerebrospinal fluid into and away from the head. Nerves and muscles also pass through and are affected by the thoracic inlet. This three-dimensional structure supports an effortless, open, wide, and upright upper chest and shoulders.

Lying parallel to and above the muscular diaphragms and thoracic inlet are the ligamentous vocal cords which, together with their vestibular folds, also give a horizontal "diaphragmatic" support at the level of the throat. In the brain, sections of the dura mater, a membranous sheath that envelops the brain, can also be felt to provide a tensile horizontal support at about the level of the ear. We can feel these latter two ligamentous and membranous structures to be supported from below by the lower two muscular diaphragms; this gives a powerful support for both the voice and the perceptions of the head. Aligning all five "diaphragms" provides a very stable but spacious and dynamic support for the internal organs and the balance of the spine, complementing the primarily vertical support of the musculoskeletal system.

Muscle Currenting and Complementary Action

Working with the balancing of eccentric and concentric contractions can facilitate the repatterning of muscle action. To further refine and repattern more deeply and specifically, we also apply the principle of muscle currenting, or sequencing of an action.

The muscles wrap around the bones, and generally their action is linear and direct, through the length of the muscle from one end to the other. Their action draws the two or more bones on which the muscle acts closer together or farther apart along the line of action. If we look at the muscular system as a whole we see that these direct lines in the individual muscles and sets of muscles combine to create a continuous spiraling movement throughout the length of the limbs and torso. If movement is initiated at the very tips of the fingers and toes and carried through to the body's center in one continual motion, the form of this movement will be spirallic as it follows the sequential line of muscles connecting from the extremities to the center. This is also true if movement begins at the center and sequences out. In development the spiral form of movement is first seen in the birthing process, but its complete development throughout the whole body occurs at the final stage of the developmental sequence when all other patterns and movement in the three basic planes have been mastered. It is seen in the Contralateral crawling pattern where movement initiates in the fingertips or toes and eyes and sequences spirallically through the muscles of the whole body to the opposite extremity.

What we may feel, if we pay careful attention to our movement, is that not only does the action sequence from one muscle to the next, but there is also a currenting of energy throughout the length of each individual muscle, not unlike the currenting of energy through an electrical wire. We can think of the currenting as a preparation for actual contraction, taking place at the small moment of initiation of movement; it can also be sensed in stillness as a sense of active direction or tone of the resting muscle. Muscle currenting occurs through the sensory-motor feedback loop of the muscle spindles, a specialized system of sensory muscle fibers found within the muscle, lying amongst the ordinary muscle cells and regulating their activity.[23] The fibers of a particular muscle do not all contract simultaneously on stimulation by the motor nerves; in the Body-Mind Centering view, contraction of muscle fibers may happen randomly, but for most efficient function it should follow a sequential flow from one end of the muscle to the other. This sequential action happens very quickly so as to feel almost simultaneous, but clear initiation and sequence of action within the individual muscles can make a considerable difference in their quality of rest and activity.

Bonnie Bainbridge Cohen proposes that there is an "ideal" direction of currenting for each muscle which does not change, whatever the action. We observe that areas of tension or weakness may be caused by muscles being currented in the reverse direction; by muscles being currented from both ends simultaneously, causing a "knotting" at the center and a "muscle-bound" quality to movement; by only part of the muscle contracting, so that some sections become overused and tense and others underused; or by a confused and disordered flow of energy through the contracting muscle cells. Clear organization at this level gives movement strength, ease, grace, and finely-tuned articulation.

Muscles, or groups of muscles, tend to work in pairs, each one of the pair lying close to the other on the same side of the joint and following an approximately parallel line of force; they perform similar but slightly different actions. Sometimes there may be more than two such muscles on one side of the joint, or only one muscle with separate bundles of fibers within it performing two slightly different functions. For simplicity, I will also refer to these variations as a "pair" of muscles. A simple movement, such as full flexion and extension of a specific joint, will primarily involve such a "pair" of muscles on each side of the joint being activated—four muscle actions in all. I will refer to this as "four" muscles, but it should be remembered that there may be more or less than four actual muscles in such a system.

In the following discussion we will be considering the group of muscles primarily responsible for carrying out a particular movement. In any given action, there are many other muscles involved that provide reinforcement or stabilization for the main activity. In fact, every muscle in the body must in some subtle way adapt to any and every movement made, supporting posture in various ways in order to maintain balance. The musculature of the body actually functions as one integrated organ of movement, changing its shape, density, and tone constantly in response to both the largest and tiniest of actions. However, as we work with specific muscles and muscle groups, we find that change in a particular area of muscles can radically affect the functioning of the whole.

In their book *Muscle Testing,* Daniels and Worthingham describe "synergistic action" as

> a contraction of all the muscles acting around a joint. These include the prime movers, the muscles that act in concert with the prime movers to define the spatial limits, and the antagonists that check or limit the movement.[24]

In traditional terms, the "prime" or "primary movers" are those muscles considered to be the most essential to the performing of a particular action; Thompson describes the primary movers as "the largest and most important muscles."[25] In the Body-Mind Centering approach to analysis of muscle action, we use this term but have redefined its use as a result of observing the different stages of an action. In the system presented here we will consider how the primary mover changes from one muscle to another throughout different phases of a movement. Sometimes the primary movers may be what is described above as "the muscles that act in concert with the prime movers," or "the antagonists." Other muscles not acting directly on the joint in question may be active as stabilizers or reinforcers but are not central to the main action. These muscles will not be included in the description below, as the aim is to focus on repatterning specific muscles and creating balanced action within clearly defined muscle groups. These secondary muscles will, however, act as primary movers of a different movement activity. They will also be indirectly affected by changes occurring in the muscles worked with directly.

Muscles complementing each other's action on opposite sides of the joint through eccentric and concentric contractions are traditionally known as the "agonist" and "antagonist."

Generally, the agonist takes the moving bone in the direction in which it is going, through a concentric or shortening contraction; the antagonist complements this by lengthening (an eccentric contraction). The traditional view holds that the antagonist lengthens through relaxation, to allow the contraction of the agonist to take place, in a process of reciprocal innervation; our view differs only in that we consider the lengthening to be active and as essential to the movement as is the concentric contraction. It is interesting to note, in terms of our culture's attitude towards movement of the body, that these terms stem from the Greek word *agonistes,* whose root meaning is "agony," and which denotes a combatant and an adversary or opposing force. The model we are using here implies an attitude of balanced interaction and complementarity, rather than one of conflict or competition. This reflects a shift in awareness and attitude that our culture needs and is finally beginning to make.

If we take just one pair of muscles, we see that they both cross over the same joint (sometimes one may pass over an additional joint) on the same side, and lie along approximately the same line of action. They will therefore be involved in eccentric or concentric contractions simultaneously. However, their action is different. Each pair will include one muscle, or group, or group of fibers, lying deep in the center of the body closest to the skeleton. These muscles tend to be the smaller and finer of the two and we classify them, for convenience, as "A" muscles. These muscles usually cross over only one joint. The other of the pair will be a larger muscle lying in the layers closest to the surface of the body; we will contact this layer first as we palpate, or sense with our hands, the muscles through the skin. These muscles are classified as "B" muscles. They often pass over two joints. (Fig. 7.17) It should be noted that the classification of muscles into two groups named "A" and "B," and the description of muscle action that follows,

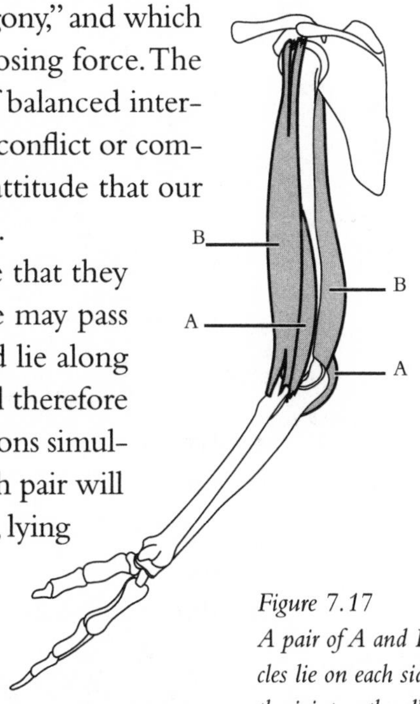

Figure 7.17
A pair of A and B muscles lie on each side of the joint—the elbow.

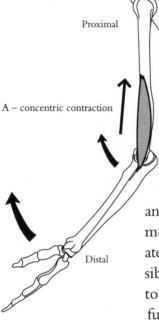

Proximal

A – concentric contraction

Distal

Figure 7.18
Proximal initiation of
movement—an A mus-
cle initiates a movement
of the distal bone at its
proximal end (flexion of
the elbow).

are original to Body-Mind Centering work. This system is derived from experiential research and practice and is related to BMC principles. Rather than conflicting with muscle theory, this approach enriches it by offering a new perspective and subtlety of definition to more traditional kinesiology theory.

The A muscles work most effectively when their action currents from the end of the muscle closest to the extremity, in toward the center of the body—that is, from the distal to the proximal end of the muscle. This sequence of action means that a movement is initiated in the distal bone, and at the joint we see the proximal end of the distal bone being moved by the muscle action. This is termed a proximally initiated movement. Therefore, A group muscles are primarily responsible for the proximal initiation of movement, by currenting in towards the center of the body. (Fig. 7.18) The A muscle will function as the primary mover, that is, the muscle most essential to carrying out the movement during the first phase of the action.

The reverse happens in the case of B muscles. Their action currents from the proximal to the distal end of the muscle, initiating a movement at the distal end of the proximal bone—hence a distally initiated movement begins with the action of a B muscle; the B muscle will be the primary mover during the initial phase of the movement. This applies whether a contraction is eccentric or concentric. (Fig. 7.19)

When the A muscles initiate and the action sequences through the whole length of the muscle, movement has a sense of active initiation, lightness, fine articulation, extension, and clarity in space. Proximal initiation usually happens in the movement of the limbs on the supporting spine, for example, in the dancer's articulation of her arms in space. The closer to the tips of the fingers, toes, tail of the spine, and muscles of the face and senses that movement begins, the more clearly and vibrantly do these qualities emerge. Distal initiation through the B muscles, where the limbs support and allow for mobility of the center of the body, gives strength, power and substance to movement. There is a sense of fully releasing energy and an ability to be deeply connected to the earth as we move. The infant learning to crawl is developing strength and coordination in the B muscles and is initiating distally; developmentally, control of the A muscles through proximally initiated movement follows that of the B muscles.

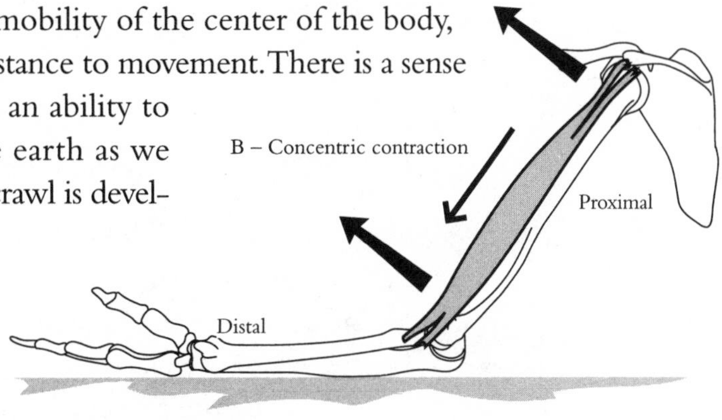

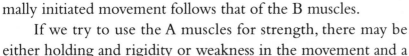

B – Concentric contraction

Proximal

Distal

*Figure 7.19
Distal initiation of movement—a B muscle initiates a movement of the proximal bone at its distal end (flexion of the elbow).*

If we try to use the A muscles for strength, there may be either holding and rigidity or weakness in the movement and a feeling of collapse. If the A muscles current in the reverse direction, spatial clarity and aliveness in the movement will be lacking; movement may appear to be fluid, but will feel limp and lifeless through the limbs. Some releasing techniques tend to encourage this quality through their emphasis on outward flow. Simply currenting the A muscles towards the center of the body, without actual movement taking place can reverse this tendency and give a sense of activation and integration.

When the B muscles are used for fine articulation and coordination, the movement will feel clumsy, heavy, and bound; currenting them toward the center will make the movement labored, effortful, and tense. It may lack balance, control, and groundedness; energy and power will appear held in, unexpressed. Cur-

TABLE 3. MOVEMENT INITIATION

	At a specific joint	*In the moving bone*	
Proximal initiation	Proximal bone supports; distal bone moves	Movement initiates at proximal end of distal bone	A muscle initiates
Distal initiation	Distal bone supports; proximal bone moves	Movement initiates at distal end of proximal bone	B muscle initiates

renting the B muscles outward, prior to actual contraction, can give a sense of release and lengthening.

Four Stages of a Muscle Action

A simple movement, such as flexing and extending at the elbow joint, should ideally involve a balanced interaction of both "pairs" of muscles, one on each side of the joint. The principal flexors of the elbow joint are the biceps brachii and the brachialis; the principal extensors are the triceps brachii and the anconeus. (Fig. 7.20) Other muscles will be involved less directly in this movement, but in order to work with the repatterning of specific muscles, we may limit our focus to these four to begin with. By redefining their respective actions, control through a fuller range of movement and a balance of strength and graceful articulation can be found.

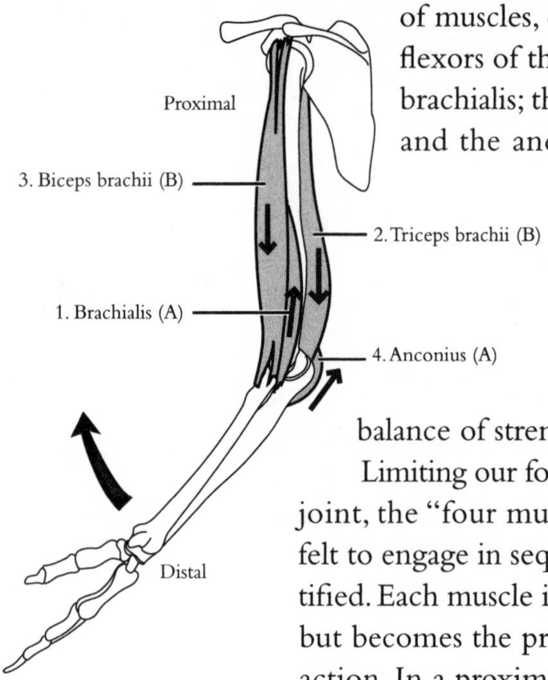

Limiting our focus to the flexion and extension of the elbow joint, the "four muscles" involved, two on either side, can be felt to engage in sequence. Four stages of an action can be identified. Each muscle is involved throughout the whole movement but becomes the primary mover during a specific phase of the action. In a proximal initiation of flexion, beginning with the arm hanging by the side or the elbow resting on a table, the sequence is as follows:

1. A concentric contraction of the A muscle which lies above the joint, relative to gravity (and hence has to pull against the force of gravity); this would be the brachialis, and possibly also some fibers of the biceps brachii;
2. The B muscle on the opposite side of the joint eccentrically contracts; this would be the triceps brachii, in particular its more superficial sections;

Proximal

3. Biceps brachii (B)

2. Triceps brachii (B)

1. Brachialis (A)

4. Anconius (A)

Distal

Figure 7.20
The four stages of muscle action—proximal initiation of flexion at the elbow joint.

3. The B muscle on the first side concentrically contracts; the biceps brachii is responsible for this, and its action takes the main force of the weight of the arm moving against gravity;

4. The A muscle on the opposite side eccentrically contracts to complete the movement; this would be the anconeus, and possibly the deeper fibers of the triceps brachii.

This last A muscle would then initiate the returning extension, and the order of involvement would reverse, as would the concentric/eccentric directions. (For further clarification, see the description in the Exploration that begins on page 176.)

For a distally initiated movement, where the forearm rests on the floor as the arm is flexed at the elbow, the pattern is reversed;

1. The B muscle on the side uppermost, i.e. the biceps brachii, concentrically contracts;

2. The A muscle on the opposite side, the anconeus, eccentrically contracts;

3. The A muscle on the first side, the brachialis, concentrically contracts;

4. The B muscle on the opposite side, the triceps brachii, eccentrically contracts.

Again the muscles engage in reverse order as the elbow joint extends, reversing concentric and eccentric.

These four stages of muscle currenting can be applied to movement at any joint or series of joints, such as those of the spine, in flexion, extension, adduction, abduction, and rotation. This at first requires careful attention, sensing the action through specific muscles and actively creating the directions through the process of thinking through the body, connecting image of movement with sensation. Once a new pattern has been established in the nervous system in this way, the conscious attention can be relaxed; the new patterning will not be lost when we cease to focus on it consciously but will continue to guide our everyday movement, creating more balance and efficiency.

To work with the muscles in a specific way, we first look at pictures of them to get an idea of their shape, location, where they are attached to the bones, and to understand their lines of action and function. Then, locating the skeletal attachments on our own or a partner's body, we sense through the layers of tissue to the depth at which the particular muscle lies, being aware of its spatial relationship to the skin, bones, organs, and other muscles in the area. Some muscles will be very easy to find and make contact with, and, being soft tissue, they can be kneaded, massaged, and gently stretched or compressed. The deeper muscles and those lying within the bony framework, or under thick layers of fatty tissue, can be located through sensing, moving and currenting in the appropriate direction along the line of action of that muscle. When a muscle is currenting clearly in its "preferred" direction, movement will feel effortless.

The "Mind" of the Muscular System

Movement and pain are the two primary ways in which we feel and locate our muscles; sensing is a more sophisticated and directive but less familiar way of contacting and perceiving them. Generally we feel our muscles in action in a less self-conscious way, through a combination of the muscular and circulatory fluid systems. (More will be said about this in Chapter Nine.) The muscles are very rich in blood and contact with them through our hands should convey this quality of weight, fullness, and fluidity. Movement initiated primarily in the skeletal muscles (not all movement is, in the sense in which Body-Mind Centering defines movement initiation), with the mind focused there and the fluids supporting, will usually express the rhythmical flow and weightedness of the blood, as well as the vitality, strength, and activity of the muscles themselves. The muscles' quality of "mind" is alive, alert, expressive, and ready to interact.

When the muscles are combined with the nervous system, the latter contributes a quality of complex organization, order, logic, and down-to-earth practicality to the "mind" of the muscular system. You might have experienced such qualities as you attempted to follow and understand the above description of muscle functioning. Whether this "mind" feels clear, difficult, rigid and dogmatic, enlivening, powerful, or frustrating will depend on your own relationship to your muscular system and the combinations of systems through which you tend to express. Combining the muscles with the nervous system will feel quite different from combining them with the fluids, bones, or organs, for example.

If the information about muscle functioning was hard for you to follow, before pursuing it further you might try reading the following chapters and exploring the qualities of attention of the different systems. In the "mind" of which systems can you most easily experience the quality of attention needed to follow this description of the muscles? Does it require a particular kind of perception and thinking? And, which systems might support this? For example, I find I need the clarity of the lymph to bring me to a place where I can follow the analytical process of the nervous system required there. The lymph helps me to accept this aspect of nervous system functioning, balancing it with a fluid quality.

By embodying a specific system, you can pattern yourself into a certain "mind" to suit the activity, or you can take a more receptive approach and choose the moment to involve yourself in that activity when you are experiencing the quality of attention required. It will also be necessary, if you are to easily follow these descriptions of muscle currenting, to actually try out the movements yourself with the help of an anatomy book to locate the specific muscles. Embodying the principles in movement will give a much clearer sense of them than will simply reading about them. The muscles need to be felt in action.

As well as needing to rest from muscular activity, we need at times to rest *in* muscular activity. If a person's work involves much complex organizational thought but little physical activity, as very often is the case in our car- and desk-bound society, he or she is likely to be engaging more the nervous system aspect of the neuromuscular system. Whole-body muscular activity, also involving the blood and other fluids, is needed to rebalance the neuromuscular system and avoid a buildup of tension and frustration. Emotional factors in the work environment also cause the release of adrenaline into the bloodstream, increasing the body's readiness to act; this creates a need to release tension in vigorous physical activity. A failure to do so can be a significant factor in many stress-related illnesses. It is equally important, when repatterning the muscular system through a sensing process, to finally let go of sensing and allow the muscles to move in a more spontaneous way in either everyday, creative, or athletic activities.

Exploration

There are several approaches to working with the muscles. A massage-like contact can be pleasurable and relaxing, as well as helpful in identifying the locations and quality of the muscles. The necessity for active and fluid movement has been mentioned above; this kind of movement is probably very familiar to you as a way of strengthening muscles, releasing tension, and to help you feel more awake, energized, and alive. The more specific repatterning techniques outlined in this section can also be explored in order to define and refine muscle action more precisely; practice of these principles can help to clarify the activity of individual muscles and muscle groups, release tensions, and strengthen and balance the system as a whole.

1. Practice with a partner making contact through your hands with the muscles. Feel their elasticity, mobility, and weight by gently massaging, kneading, rolling, lifting, lengthening, and

compressing. Feel that you are working from the muscles of your own body, balancing the qualities of weight and fullness with fluidity and elasticity in your touch.

2. Using anatomy books to identify and locate specific muscles, try to determine whether they are A or B group muscles. You can locate in your own or a partner's body the origin and insertion (the two ends) of a particular muscle and trace its line of action, to guide your partner or yourself into currenting through eccentric and concentric contractions, along the whole length of the muscle.

For example, look at a picture of the muscles of the neck and shoulder and find the levator scapulae and the upper portion of the trapezius muscle. These two function as a "pair"; the levator scapulae is an A muscle and initiates the raising of the scapula up towards the base of the back of the skull (a proximal initiation); the upper trapezius is principally a B muscle and, when its insertion is fixed, initiates the drawing of the base of the skull down and around toward the outer tip of the shoulder (a distal initiation). Current these two muscles through eccentric and concentric contractions. This is an area of tension and lack of differentiation in many people; clear currenting of these muscles can greatly help to release holding patterns. (Fig. 7.21)

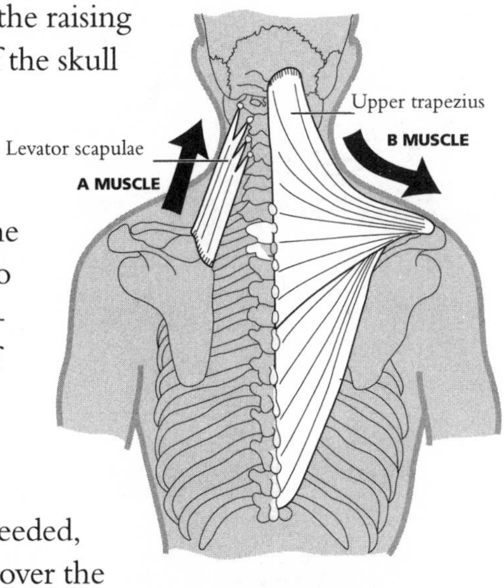

Figure 7.21
Currenting of the levator scapulae and upper trapezius muscles.

3. See if you can feel, or repattern where needed, the four stages of action in the muscles running over the front and back of the knee joint. Focus on the four muscles of the quadriceps femoris at the front of the thigh, and the three muscles of the "hamstrings" at the back. These muscles attach to the top of the tibia at their lower (distal) ends, and on either the pelvic bones or upper femur at their upper (proximal) ends, hence forming the bulk of the large muscles of the front and back of the thigh. See if you can identify the A and B muscle groups.

For the proximal initiation, lie on your front and slowly raise the foreleg (lower leg) off the floor and up towards the back of the pelvis, beginning with:

I.   A concentric contraction of the deep A muscle currenting up over the back of the knee, (semimembranosus and deep fibers of biceps femoris),

II.   An eccentric contraction of the superficial B muscle down the front (rectus femoris),

III. A concentric contraction of the superficial B muscle currenting down the back (semitendinosus and biceps femoris), and ending with

IV.   An eccentric contraction of the deep A muscle currenting up the front (vastus lateralis, vastus intermedius, and vastus medialis).

Reverse the order of muscles, and eccentric or concentric contractions, to return the foreleg to the floor:

I.   A concentric contraction of the deep A muscle up the front of the thigh,

II.   An eccentric contraction of the superficial B muscle down the back,

III. A concentric contraction of the superficial B muscle down the front, and

IV.   An eccentric contraction of the deep A muscle up the back.

Practice slowly several times, and notice if there are any changes, particularly in the pelvis and the knee joint itself. (Fig. 7.22)

To initiate distally at the knee joint, sit back on your heels with your hands on the floor in front of you. Slowly shift your weight forwards onto the hands, so that the knee joint gradually extends, beginning with:

(A) Flexion

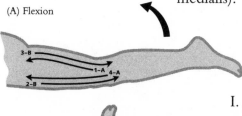

(B) Extension

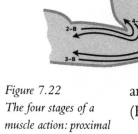

Figure 7.22
The four stages of a muscle action: proximal initiation of (A) flexion at the knee joint; (B) extension at the knee joint.

   I. A concentric contraction of the superficial B muscle running down over the front of the thigh and knee (rectus
femoris),

  II. An eccentric contraction of the deep A muscle up the back
(semimembranosus and deep biceps femoris),

III. A concentric contraction of the deep A muscle up the front
(vastus lateralis, vastus intermedius, and vastus medialis), and
finally

IV. An eccentric contraction of the superficial B muscle down
the back (semitendinosus and biceps femoris); the knee joint
will now be more fully extended.

To flex the knee again, and rock back to sitting on the heels,
reverse the order:

   I. A concentric contraction of the superficial B muscle down the back,

  II. An eccentric contraction of the deep A
muscle up the front,

III. A concentric contraction of the deep A
muscle up the back, and

IV. An eccentric contraction of the superficial B muscle down the front.

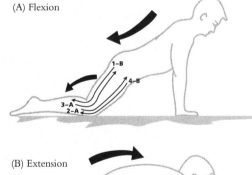

(A) Flexion

A contraction is eccentric or concentric according to whether the two bones are
moving apart or towards each other on that
side of the joint. (Fig. 7.23)

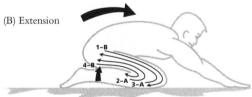

(B) Extension

(Be aware that muscle groups in other
areas of the body will also be engaged during
these movements; a similar sequencing of muscle action will be
occurring at the hip, shoulder, and wrist joints in the movement
above. The principles of the "four stages" can also be applied to
muscles at these joints, but focus on what is happening at only
one joint at a time, to avoid confusion.)

Figure 7.23
*The four stages of a
muscle action: distal initiation of (A) flexion at
the knee joint; (B)
extension at the knee
joint.*

After practicing several times slowly, stand and walk naturally, noticing whether you feel any changes in your posture or walking. The walking also helps to integrate any changes that may have taken place.

4. Improvise freely with your movement, paying attention to exercising all of the muscles you are aware of, alternately lengthening out and compressing in, to feel the rhythmical nature and vigor of muscular activity. Remember to include the muscles of the hands, feet, face, and sense organs. Music or the presence of other "dancers" can help to free the sensing process and bring fluidity and spontaneity to the muscles. Explore both strong and delicately articulated movements, supporting and pressing the limbs against the floor and moving them freely in space. Be aware of the heat that this generates, and notice the quality of attention, perception, and communication. It is through the muscles that we actively participate in and interact with the environment.

Chapter Eight

The Contents:
Soft Tissues of the Body

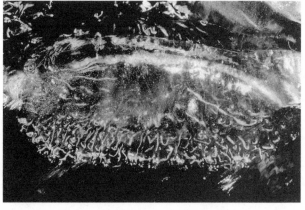

Within the musculoskeletal framework are contained the soft tissues of the internal organs, or viscera, that carry out the vital work of maintaining, renewing, and reproducing life. Several physiological systems are involved within the organ system as a whole: the digestive system, the respiratory system, the circulatory system, the lymphatic system, the reproductive system, the urinary system, and the nervous system. The brain and special sense organs are classified within both the organ and nervous systems.

There are also a number of different types of glands, falling into two basic categories: "exocrine" and "endocrine" glands. Exocrine glands, such as sweat glands, salivary glands, mammary glands, and those involved in digestive processes, secrete their fluids through ducts and thus affect a specific organ or function of the body directly. Endocrine glands are ductless and secrete hormones directly into the bloodstream, affecting the person as a whole in both a specific and a general way, physiologically and emotionally. Some organs, such as the ovaries, also function as endocrine glands; others, such as the pancreas, contain patches of both exocrine and endocrine tissue and function as both types of glands.

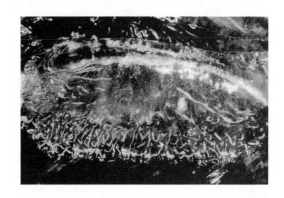

The Organ System

The organs have volume and weight and fill the area within the torso and skull. They give fullness and presence to the body and its movement—aliveness and expression of feeling. (Fig. 8.1) Each organ's activity generates or supports a different state of mind and quality of expression in movement that reflect its activity, structure, substance, size, and position in the body. Similarly, expression or repression of emotion will not only reflect the organ activity but also affect the organs through stimulation or inhibition of their full functioning. When we laugh, cry, or explode with anger, we experience this *e-motion*—a movement outward—in the organs. Reich, Lowen, and Dychtwald, among other body-mind therapists, have carried out extensive work and research into the connection between the emotions and the functioning of the organs. Their findings have relevance to our present study.[1]

The experience of the newborn infant is primarily organic—a subjective involvement in the sensations of feeding, digesting, eliminating, breathing, movement, and sensory stimulation. Emotional life develops and a primitive sense of identity begins to form on the basis of the sensations stimulated by these processes, which are perceived as pleasurable, painful, and so on. The organs connect us to our "gut" feelings and reactions: raw, unbounded, and uncensored emotions. They can evoke or

Figure 8.1
The organs give aliveness and expression of feeling to movement.

183

reflect the pleasurably powerful state of fullness of being, the frustration of being unable to let go, the pain and fear of being empty or abandoned, the joy of being alive, and so on. The organs evoke and express the sense of differentiated being in its myriad of feeling qualities.

The Supportive Function of the Organs

In addition to their physiological functions, the organs serve an important role in the support of bodily movement and posture. Each organ is supported within itself by a "skeleton" of fine connective tissue that branches throughout its structure and by the full internal respiration of the cells, which make the organ as a whole alive to its own presence. Being alive to its own presence means being aware at the cellular level; as its cells breathe fully the organ will expand into its own space. In this way each organ energetically supports itself and the organs surrounding it. The fullness and aliveness of the organs also give support to the musculoskeletal framework from within. Imagine a deflated balloon: it collapses in around itself. If the balloon is filled with air it takes on shape and volume and is supported by the air inside; it becomes an entirely different object.

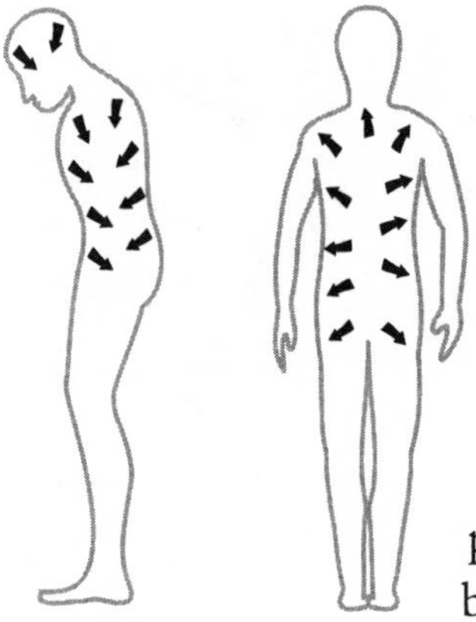

In the same way, if the body's framework is not supported from within by the soft tissues, it will tend to fall in around its center. The organs collapse away from the musculoskeletal wall and will be felt as a dead weight to be carried around passively. (Fig. 8.2) This puts stress on the joints and muscles, which causes movement to feel restricted and heavy, while posture loses its internal support. The organs may be felt as a burden rather than a source of power and vitality. Or there may

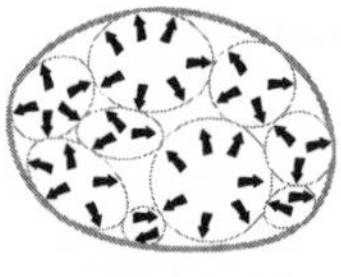

Figure 8.2
The organs, when active, support the musculoskeletal framework from within. When they don't support, the framework tends to collapse in and downward, as the organs draw away from the body walls.

184

be a sense of emptiness and insubstantiality that can create feelings of ungroundedness or unreality. We find that each organ gives support to specific bones, joints, and muscles, although each has relationship to every part and can be brought into a supportive alignment with other organs, bones, or muscles according to the unique needs of each individual's overall patterning. For example, we experience the kidneys as having a direct energetic connection to the knees, and misalignment or stress in one area may be reflected in the other. In working with a knee problem we might look at the alignment of the knees with the kidneys and the way the kidneys are or are not supporting the lower back. Lack of organic support for the lower back will cause weakness there and a tendency for the lumbar region to either collapse or hyperextend in sitting and standing. In either case, the pelvis is not free to transfer the weight of the body into the legs through a clear and balanced alignment; stress will be thrown into the knees, which tend to lock or hyperextend to maintain balance. For a particular individual suffering from lack of support in the kidneys, it may also be helpful to suggest a feeling of tensile support or countermovement between the kidneys and the elbows, sternum, head, or heart, for example, according to the different needs of each person.

We experience the heart as having a special relationship to the hands and eyes in terms of mutual energetic support and expression. Heart energy is expressed through the eyes and the touch of the hands. When we experience the heart as full and present, without fear, we are able to reach out through the eyes and hands to give and receive. As the hands and eyes meet with the environment in this way, supported from within, the heart is also supported, sustained, and nourished by the world outside—the support flows in and out. But if the heart withdraws and collapses, it does not support nor is it supported by the hands and the eyes, and the free flow of energy is disturbed. A lack of tensile support and dynamic connectedness between the heart, hands, and eyes will result in a collapsed posture and a broken connection between the inner and outer worlds. A painter friend of mine once described how essential this three-way connection between heart, eye, and hand was to the flow of his artistic creativity.

Balancing the heart with the brain or the uterus, for example, might also give a greater sense of inner integrity and support; there is also a suggestion here of bringing into relationship and balancing the energies of the heart, thinking, and sexuality. We could explore the relationship of each organ to any other organ or area of the musculoskeletal framework to discover which connections we personally need to make in order to create a feeling of balance and integration.

The energy of the heart is expressed through the hands and also provides support for them as they move in space. The energy of the organs needs to be allowed to reach through the limbs and out into space; to do this we need to go through another system, such as the bones, muscles, blood, or nerves. The heart also supports the upper body, spine, and head when weight is taken on the forearms and hands, as in the "sphinx" posture. The weight of the heart is transferred through the bones to the hands, and its energy radiates out through the hands. As the hands push into the floor, this impulse levers back into the heart to give support to the spine at this level. This also helps to lift the head. Gravity is not just a downward force but also supports upward and works with us to support us. (Fig. 8.3)

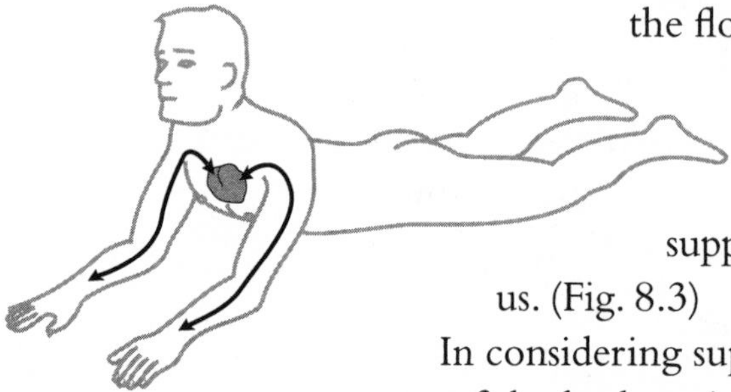

Figure 8.3
The energetic connection between the hands and the heart gives support to the upper spine and head.

In considering support we are dealing with the weight and mass of the body as it lifts itself upward against gravity. In the upright posture the actual weight of the body should not fall through the organs in its center, which will cause them to become overcompressed and collapsed. The unnatural strain to which they are subject may cause a variety of organic disorders. Actual weight should be transferred through the bones, the densest tissues of the body, which are best designed for this purpose. Each organ has its own volume that supports itself and the container energetically, but its substantial weight is transferred through the skeleton to the ground. However, "the weight is balanced around the body's axis which passes primarily through the space of the body cavities"[2] in which the organs lie. This can happen when the organs are active and well-toned and are therefore able to support their own space energetically.

When energetically present and alive, the organs can act as a support for posture and movement in two ways: through compression, where the weight of the body, or body part, presses through them into the floor; or through suspension, where the body weight suspends from the organs. For example, in the shoulder stand yoga pose, the weight of the body falls into the organs of the upper chest and throat, thus stimulating their tone. If these organs are energetically active in this position they give a compressive support for the rest of the body as it lifts up out of gravity. The organs that are not in contact with the ground, primarily those of the abdomen and pelvis, give a suspension support if they are active; the weight of the upper body is felt to suspend from the pelvic organs and be supported by them. (Fig. 8.4) The "lift" experienced through the pelvic organs can be extended energetically through the legs so that they too will lift away from the upper body and feel supported and light. If the organs do not actively give this suspension support, the weight of both organs and legs will sit heavily into the shoulders and neck.

The reverse of this would be the "sphinx" posture, or the cobra pose in yoga. Here the organs of the abdominal and pelvic regions, as they are lying in direct contact with the ground so that the body weight passes through them, give compressive support. The organs of the chest, throat, and head are lifted off the ground and give suspension support. It is important to balance the activity of the organs throughout the body by varying their

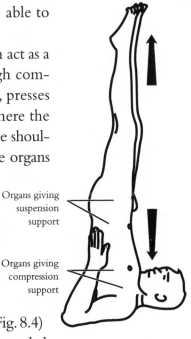

Organs giving suspension support

Organs giving compression support

Figure 8.4 In the shoulder stand, the organs of the upper torso and throat, when active, give a compressive support, as the weight of the body falls through them; those of the lower body give suspension support, as the body weight is felt to suspend from them. In normal upright posture, this is reversed.

functions as compressive and suspension supports. Hatha yoga offers an excellent method of doing this.

The organs can also energetically support the limbs and their movement. This can be facilitated by feeding a compressive force in through the limbs, as described in relation to the Push patterns and work with the skeletal system, but this time focusing on directing the energy through the limbs into the organs themselves. When the energetic connection is made there will be a feeling of buoyancy, a dynamic rebound quality, as the energy of the organs begins to flow outward through the limbs.

The sensation of supporting in this way from within is very different from that of muscularly holding up out of gravity. There is a direct relationship between the qualities of organic and muscular tone. Where the organs are not providing support, either the muscles will have to work harder and areas of tension will develop in order to maintain posture, or else there will be a weakening and collapse of the structure as a whole and a general lack of alertness and vitality will result. When organic support is lacking, movement will show strain. Good balance comes from inner support, and this creates strength. Support and balance within the body are a process of continuous cellular awareness, not a static goal to be achieved. This is true also of emotional experience, support, balance, and strength, which develop out of and are closely related to the organs' physical processes. These are aspects of our life's work, not something to be quickly found and dispensed with. Joseph Campbell describes myth as an organic process—the ongoing and ageless story of our psychological life.[3]

Activating the Support of the Organs

To begin to contact and activate the support of the organs, excessive muscular tension in the area may first need to be released. To facilitate this letting go, it is helpful to work first in a lying-down position so that the whole-body weight is supported by the ground and muscular tension is not required to maintain posture.

Alternately, a well supported sitting position could be used. Sometimes it can be helpful to relax excessively tense muscles with massage or cellular breathing before working directly on the organs. Visualizing the organs, their shapes, locations, size in all dimensions, and relationships to each other, to the skeleton, and to the muscles, helps to bring the attention to the sensation of their presence. Then we imagine breathing into a specific organ, feeling it expand equally in all directions from its center during the inhalation. On the exhalation, imagine lightly maintaining this volume without straining to do so. If breathing in this way creates stress, it is helpful to feel a sense of allowing the breath to come from the organ rather than forcing the breath into the organ. This facilitates the internal respiration of the cells of the organ.

Particular attention is given to breathing into any area of an organ that feels lifeless, contracted, dark and heavy, or immobile. We feel for the quality of tone in the organ: what the sense of energy and activation is, where the weight is most concentrated, how free the flow of energy and movement is, what the organ needs. We can easily train ourselves to sense this information and use our active imagination to awaken full cellular breathing within every area of the organ. It will begin to feel more full, perhaps weighted, or light and weighted at the same time, and the muscles and other tissues surrounding will begin to relax to accom-

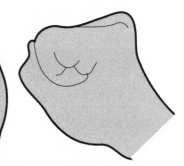

modate the inner expansion as the organ finds its own volume and support. Breathing in such a way as to facilitate their expansion is helpful for those organs that feel too tight, contracted, or hypertoned. The force or energy of a hypertoned organ moves inward to the center; a feeling of expansion and movement outward in all directions from the center needs to be created. (Fig. 8.5)

If an organ already feels too expanded, flaccid, or unintegrated, it may be hypotoned. In this case the energy is felt to move

Figure 8.5
An organ may feel too contracted, like a tight fist; the energy or forces within the organ need to be directed outward to allow for softening and expansion.

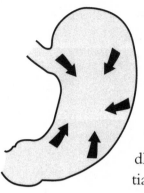

Figure 8.6
If an organ feels too expanded or lacks integration, the energy needs to be directed in toward the center. The digestive organs in particular often suffer from this tendency.

outward, and the forces need to be brought together. We can visualize and actively create the sensation of knitting together around the center of the organ as we breathe into it (Fig. 8.6); or we may sense the surrounding tissues and organs holding it and limiting its expansion. For example, the heart expands within the containment of the surrounding lungs. We can think of holding and rocking the heart within the "cradle" of the lungs, so that the expansion of the heart's potentially infinite energy may also be supported and contained. The imagery that we use can be created and adapted to suit individual needs. As we learn to make contact with the organs and listen to their needs, we can actively create the conditions that will encourage their full and healthy functioning. This process of active imagination gradually aligns with and awakens awareness of the actual feelings and sensations present. As this happens new sensations and movement patterns are consciously created.

Sounding or toning into the organs is also a powerful means of enlivening them and giving expression to their energy. Each organ has a unique vibration of energy, and when we train ourselves to be sensitive to their distinctions we can differentiate one organ from another. Through listening inwardly we can "hear" the organ's "tone," the vibration at which it resonates. Sounding this tone helps awaken awareness and stimulates the energy of the organ, just as the organ stimulates and supports the sound. There is a mutual support between the organs and vocalized sound, through which we give expression to our inner feeling world in speaking, singing, shouting, laughing, and so on. Without the support of the organs, the voice lacks feeling, depth, and resonance; with this support the whole body becomes the resonating chamber of the voice and is enlivened and empowered by it.

The hissing breath is another way to make contact and activate support, and is particularly helpful in countering sluggishness in the organs. By partially closing the teeth and mouth cavity and letting the breath out in either one long sustained hiss or a series of short, clear, and rhythmical hissing sounds, we set up a counterthrust into the organ where the attention is directed. This has the effect of toning and strengthening as well as expanding the organ. The hiss should continue only for as long as a feeling of expansion or activity in the organ can be maintained; again, this should not be forced or stress may result. The organs respond best to gentle and subtle work. You may find different organs will respond more readily to breathing, sounding, or hissing.

The tone of the organs is also stimulated by the pressure of weight falling through them. As the newborn infant first begins to lift its head, weight falls through the organs of the throat; this area acts as a "keystone" for this posture. The organs of the throat are stimulated by the compression caused by the weight of the lifted head. Compression stimulates the organs to move, and so the neck becomes more mobile. The infant is then able to lift its head higher and weight falls through the organs of the upper chest, which now become the base of support. This process continues down through the whole torso to the organs of the pelvis, as the head is gradually lifted higher and higher from the ground. The organs are activated as the weight falls through them in this way; wherever the stress is located, this is where the organs are developing. When all the organs have been stimulated through compression in the horizontal position, they are able to support themselves and are ready to support and initiate movement in different relationships to gravity.

Great stress can be created if weight falls habitually through certain organs, or parts of organs, in one particular direction only. If an organ is being continually used as a base of compressive support, it loses its tone and its ability to move freely. Good tone means the ability to change; a hyper- or hypotoned organ becomes inflexible and unable to respond. With this kind of habitual pattern, there will also be other organs or organ surfaces that rarely experience themselves as the base of support. Lacking the stimulation of the compressive force of weight falling through them, they too lose their tone and the ability to support. In this way organs may become too tense and compressed, or too weak and flaccid. In both cases, stress can result.

If we move and frequently change our position the organs experience themselves in different relationships to gravity, to each other, and to the body as a whole. The different organs and organ surfaces then alternate between being supporting and moving structures. "The underside of an organ is the supporting surface,"[4] and it is here that change can occur. As the body moves, the organs rotate and a new surface becomes the support. This increases proprioceptive stimulation. Rolling the body in all directions enables every organ and organ surface to experience itself as the base of support, and so to be toned and strengthened. As we roll the weight shifts, like sand or water in a container, to the new underside, meeting the force of gravity and bonding with the earth. (Fig. 8.7) The upper side becomes empty of weight, light and mobile, as the underside becomes weighted, grounded, and supportive. This dynamic interplay, a dance between weight and lightness, gravity and levity, support and mobility, is made possible through the movement of the organs. Each organ can take on both the stable-supportive and mobile-expressive functions, and there is a continual interchange of these roles between them as movement happens.

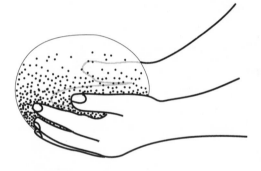

Figure 8.7
The weight of the organ shifts, like sand, to the underside as we change positions.

Repatterning Movement through the Organ System

Bonnie Bainbridge Cohen and coauthors describe the principle of relationship between organs and movement initiation in this way: "Organ support precedes initiation of breath, which precedes movement."[5] This underlying support gives fullness and power to both movement and voice. Using this support we can consciously initiate movement from the organs, which further tones and energizes them. By nature, they are not static but constantly moving in the process of their physiological functioning, each with its own rhythms and directions. The diaphragm and lungs in breathing, the heart in pumping blood, the stomach and intestines in digesting and passing food through the body, the bladder in retaining and releasing fluids, the ovaries and uterus in their cycles of ovulation and menstruation—all are pulsing at their own rates. These rhythms underlie the rhythms of our movement and emotional expression.[6] There is also movement between the cells of the organs in response to the impulses of chemical and hormonal processes, the movement of other cells, messages from the nervous system, and the activity and requirements of the organ and the organism as a whole.

Movement of the organs also happens in response to bodily sensations—pain, comfort, heat, or cold—and to the complexity of emotions constantly moving through us that in part develop out of such sensations. Often we are not consciously aware of an emotion within us, but it still affects and moves the organs and is registered unconsciously through the nervous system. The movement of traumatic or prolonged emotion that is not released or channeled into outward expression and transformed can pull or torque the organs out of place. This tension creates chronic holding patterns within the organs that underlie many of the postural problems we see in the musculoskeletal structure. If such holding patterns are organic in origin, they will be expressed through our mental and emotional attitudes.

The torquing of the organs has further implications for our perceptual and psychological development. As Bonnie Bainbridge Cohen writes, "If within the same organ and/or among different organs, there is a torque, each part registers a different place and direction which leads to confusion as to where the 'total' person is in space."[7] To release such a postural holding pattern in the body structure, attention needs first to be given to releasing the movement of the organs so that they can return to their natural position and support a better alignment.

In repatterning movement through the organ system we are utilizing this natural ability of the organs to move in response to stimulation from conscious or unconscious thoughts and feelings. Through focused awareness of the breath, which is a bridge between conscious and unconscious processes, we can actively perceive the location and state of the organs and initiate the desired pattern that will free their energy and redirect their movement. Sensory nerves present in the organs respond to sensations of hunger, thirst, pain, comfort or discomfort, tiredness or arousal. They can also be trained to give information as to the general condition of the organ, its position in space, and movement within the body.[8] As we become sensitive to perceiving this information through practice, the feedback assists our ability to consciously and actively direct the movement of the organs.

A sense of integration needs to be felt before initiating movement in an organ. The idea of "knitting together" around the center can be used, as described earlier, if an organ does not feel well integrated. Each organ can be rotated around each of its three axes in the three basic planes of movement. The ease or difficulty felt in doing this will show whether the movement is held or restricted in a particular direction or the organ habitually pulled away from its central alignment. To help free such a restriction, first imagine the direction of rotation and its axis through the organ, then follow this movement in your body, rotating the organ as far as you can in the preferred direction; finally release the movement into the opposite, less free direction as you exhale, letting the image of the rotation carry the movement to its fullest extent. (Fig. 8.8) We use the intent of the mind or creative imagination, rather than the force of will, to effect such change—the imagination has a greater power to effect change in the body than does a forceful use of the will.

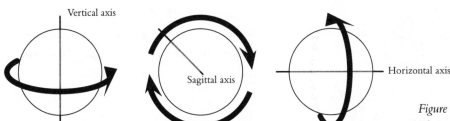

Figure 8.8
The organs can be rotated in three basic planes of movement.

The principles of a countermovement as support, or two counterrotating cogwheels, can be applied to organs as well as skeletal joints: "to achieve movement in any plane an organ can be counterrotated within its parts, with other organs, or with surrounding tissues, bones, muscles or air."[9] This strengthens individual organs and helps to free the "gluing" that often occurs within the fascia between organs, muscles, and bones, thus allowing them to move freely and independently of one another rather than being bound in an unnatural unit. This freedom allows for fluid interaction and relationship between the parts, a process that we also discern in the developing psyche of an individual or among the individuals of a community. The microcosm of the inner body will always reflect the macrocosm of the world that the individual inhabits, reflects, and creates.

Once we are able to perceive and direct the initiation of movement in the organs, we can extend this subtle action into the musculoskeletal framework by releasing the whole body into the direction of movement initiated by the organs. The organs carry their container into a fuller and rounder range of movement; as they move, habitual patterns of muscular initiation and inhibition are released. We find that specific organs will initiate and support the movement of particular areas of the musculoskeletal framework more readily in certain planes and dimensions because of their positions in the body, their shapes and sizes, and the angles at which they lie. The organs' size, weight, and rhythm will also affect the quality of activity expressed through them. (Fig. 8.9)

Figure 8.9
The initiation of move-
ment in the lungs is
released through the rib
cage and arm.

Once we sense that we have initiated movement in an organ, it is important to let go of the careful intending and sensing process. The energy of the organs needs to be released into full and spontaneous movement and the mind must also be able to move. At this point we simply move and feel the movement.

Qualities of the "Mind" of the Organs

Through their physiological functions (respiration, digestion, etc.), flow of energy, rhythm, movement, and actual physical connections to one another, the organs form systems within a system. Through contemplating their structure, function, and movement we can gain some idea of the emotional attitudes that individual organs and systems of organs reflect and support. Each organ embodies a polarity such as acceptance and rejection, love and fear or hatred, courage and timidity, joy and anger, sadness and sympathy. When we bring our awareness to a particular organ we may experience such feelings; we may also be able to perceive the relationship and attitude we have toward those feelings and the organ itself. In health, expression of both aspects of the polarity should be available to us. If there is an excess or deficiency of either, then there is an imbalance in an organ's expression that will affect the psychological and physical functioning of the person as a whole.

Such information is used in the diagnosis and treatment of illness in classical Eastern healing traditions such as acupuncture. Fundamental to these traditions is the belief that each organ corresponds to one of the five elements of nature and is similarly associated with a particular emotion, sound of voice, color, smell, season of the year, time of day, dream symbol, and so on.[10] In *The Healing Power of Illness,* Dethlefsen and Dahlke also explore the connection between mind and body in health and sickness.[11] They describe symptoms in the organ as signals pointing to a state of illness in the person as a whole; in their view, the state of illness is caused by a psychological imbalance and is expressed in the organs which reflect the function which is disturbed. Their approach, to my mind, is rather too prescriptive, but the basic principles of their work are useful. The art and science of homeopathy also considers physical and psychological symptoms to be integrally related, and treatment will affect all levels of the patient's being. The kind of symptoms that an individual displays will be directly related to their basic constitution and to their psychophysical qualities, strengths, and vulnerabilities.[12] The homeopathic view holds that our problems are intimately related to our unique strengths and qualities, as Bonnie Bainbridge Cohen also teaches.

Below are some suggestions as to the relationships between organs and feeling states. These, however, should be taken only as general guidelines. Such generalities cannot contain all the varieties of individual experience and response. You are strongly encouraged to listen to what your own body has to tell you; information received in this way will hold the most meaning for you. A valuable approach to working with an organ in distress is to dialogue with it, to inquire into how the organ itself feels, what are its attitudes toward life, its wants and needs. Remember that each organ has an important and vital role to play in the health of the whole person; and, as reflections of different aspects of our totality, each has a voice that needs to be heard and a valid message to impart to us. Processes of inner dialogue and subpersonality work, widely used in psychotherapy, can be aptly applied to learning about what our organs are attempting to tell us through their physical symptoms.[13] Creative approaches such as writing, drawing, working with clay, or expressive movement can greatly enhance such explorations.

Organs of the Digestive System

The digestive tract is essentially an open-ended tube that runs from mouth to anus. Although it is within the body, its space is continuous with the space outside. It brings the external environment within and releases it out again. The quality of its functioning tends to reflect many of our attitudes towards nourishment at all levels of existence— physical, emotional, mental, and spiritual. How we accept or reject, digest, assimilate, integrate, choose (what is to be retained and what eliminated), and let go in terms of food, nurturance, material possessions, people, ideas, and so forth, are attitudes of the digestive process. Whether we enjoy our food, attack it, play with it, dislike it, or give it very little attention, may all be signs of our underlying attitude towards nourishment and nurturing and ultimately the meaning and value of life itself.

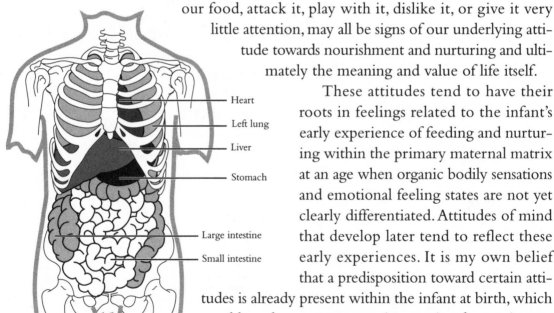

Heart
Left lung
Liver
Stomach
Large intestine
Small intestine

These attitudes tend to have their roots in feelings related to the infant's early experience of feeding and nurturing within the primary maternal matrix at an age when organic bodily sensations and emotional feeling states are not yet clearly differentiated. Attitudes of mind that develop later tend to reflect these early experiences. It is my own belief that a predisposition toward certain attitudes is already present within the infant at birth, which would tend to recreate certain emotional experiences. However we view this philosophical question, we still see a relationship between physical experience, emotional feelings, and mental attitudes. These may continue

The organs of the torso.

to interact in negative cycles until we can become conscious enough of the patterns to be able to make new choices.

The liver, gall bladder, and pancreas, while not part of the digestive tract itself, are part of the digestive system and fulfill many important functions in the process of digestion and regulation of metabolism. The liver in particular fulfills many varied tasks in the synthesis, storage, and distribution of nutrients to the rest of the body. Chinese medicine considers it to be the "general" of the body, holding an enormous amount of decision-making responsibility for the regulation and maintenance of chemical balance within the internal milieu of the body. Nutrients from the digestive tract are first sent directly to the liver to be processed before being circulated to the rest of the body as and when they are required. The pancreas and gall bladder secrete enzymes necessary to the digestion of food into the digestive tract. These organs express the maintaining of balance and control, the survival of life at a physiological level, and, when overburdened by their multitude of responsibilities, may give rise to feelings of anger and frustration or collapse and despair. They are closely associated with the solar plexus, the center of the "I," or ego, and tend to reflect issues of personal survival and the needs of the ego in relation to its place in the world.

Organs of the Lymphatic System

The spleen also performs several functions. It is an organ of the lymphatic system (this will be discussed more fully in Chapter Nine in the section on fluids); as such it produces white blood cells, or antibodies, and filters the lymph in the same way that the lymph nodes do. The spleen is therefore an important organ in the system of defense against infection and disease. Old red blood cells are also broken down in the spleen, and reusable elements, in particular iron, are saved to be used in the making of new blood cells, while other wastes are excreted. The liver and bone marrow also share these functions of production of white blood cells and the breakdown of old red blood cells. Along with the skin and lungs, the spleen stores blood until such time as it is needed, for example when there is severe bleeding; the organ then constricts to release the blood to the vital parts of the body. Because many of the spleen's functions are also performed by other organs and tissues, and because its unique role is not well understood, many Western practitioners maintain that removing it creates no significant ill effects. (An entire generation lost their tonsils—now more highly valued—by a similar rationale.) However, no other organ breaks down red blood cells in just the way the spleen does nor stores blood in the same way.

Both energetically and psychologically, the spleen is closely associated with protective and nutritive functions; the quality of this organ can relate to the general condition of the immune system, the general state of vitality of the individual, and feelings of rootedness and home. In Chinese medicine, it is considered the "mother" of the body. The fact that in the West its surgical removal is not considered to be particularly problematic seems to reflect our cultural difficulties with finding a right relationship to issues of defense, nurturance, earth, and home. Too often, medical practice intervenes with the body's natural processes of defense and healing, and the organs concerned with these functions further lose significance in our culture's concept of the body. This attitude will inevitably lead to further weakening of these functions and of the organs concerned with them. A vicious cycle between attitudes and practice is thus established and then perpetuated.

Organs of the Urinary System

The kidneys and bladder regulate the body's fluids, determining what is to be retained or expelled and when, through processes of breakdown, purification, and elimination of wastes. An enormous amount of blood passes through the kidneys daily. Their function of maintaining the healthy flow of the body's fluids and the right balance of water and minerals within the blood is vital to life. These organs are considered to be the storehouse of the "Vital Essence," or life-force, in Chinese medicine. They are associated with vitality, courage, and commitment; in a weakened state, they are associated with lifelessness, exhaustion, fear, and stress. Many people tend to suffer from stress or exhaustion in the kidney area, reflecting our cultural difficulty in knowing how to nurture this vital life force and maintain the right internal balance amid the pace and pressures of modern life. Reflecting its function of alternately holding and releasing, the bladder might be associated with feelings of buoyant support, containment, and directionality. The kidneys and bladder can also be adversely affected by a misaligned or poorly supported spine. The resulting poor postural and movement habits can put pressure on these organs.

Organs of the Respiratory System

Respiration also links our inner and outer environment in a continuous stream of breath. The lungs, diaphragm, and respiratory tracts are the organs of respiration. The breath is thought to carry *prana, ch'i,* or vital energy. It is also thought to carry the stream of thought: quieting the breath stills the mind—the essence of meditation. Breathing is also a gateway between conscious and unconscious processes. The lungs themselves reflect sadness, grief, sympathy for others, and new hope. Through the process of breathing we inspire and expire, draw in energy for new life and creativity, and release that life from us into death. Each exhalation is a loss, a letting go, but also a gift to the plant kingdom fed by the carbon dioxide we expel. Each inspiration is a return gift from the plants that produce the oxygen we need to renew and sustain our life. The interdependence of living systems and the cycles of life, death, and rebirth are reflected in the process of breathing. In each moment there is death and new life, new thought, idea, feeling. Each act of creativity has its conception in inspiration, a pause—the space in which potential is manifest—and a birth/death as the created act is "expired" and allowed to emerge. When we don't breathe fully in or out, we are withholding ourselves from full participation in both our living and dying, receiving and giving, as they express themselves from moment to moment.

Organs of the Reproductive System

The reproductive system, consisting of the uterus, ovaries, fallopian tubes, vagina, and clitoris in women, and in men the testes, seminal vesicles, prostate gland, and penis, concerns physical creativity and the expression of sexuality. It insures the continuation of the species at a biological level through reproduction, and provides emotional experiences of pleasure, fulfillment, and union with another person. Personal power or charisma, intimate relationship, the physical expression of love, and ecstasy are experiences of sexual energy. The "mind" of orgasm is also closely associated with the "mind" of enlightenment, as boundaries give way to union. The mastery and the mystery of sexual energy is followed in various mystical traditions as a path to spiritual realization. Sexuality expresses a mind both deeply personal and universal, concerning self-preservation and pleasure as well as the shared concern for the physical survival and spiritual unity of humanity. The gratification or transcendence of personal needs are the creative urges felt at work through this body system. Through it our instinctual "animal" nature can come into relationship with our conscious "divine" nature.

Organs of the Circulatory System

The heart has the mediating function in this meeting of our earthly and heavenly natures. Through it, love and compassion—

for ourselves, others, and for life itself—are felt and expressed. Most of us feel the pain of old wounds and brokenness surrounding this inner core, and the layers of defense we have created in order to protect ourselves from what we experience as the fear and vulnerability of our soft and sensitive heart. The heart expresses both our human sensibility and our potential for compassion and wholeness. It concerns sharing in a deep sense, the giving of and receiving into ourselves. When the heart is able to express fully it is not in fact vulnerable, as we so often perceive it, but extremely powerful.

The heart organ is a powerful muscle that pumps life-sustaining blood throughout the whole body. It is the central organ of the circulatory system. A special artery, the "coronary" artery, the vessels of which literally "crown" the top of the heart, carries freshly oxygenated blood first to the heart itself. The heart first nourishes itself so that it may then nourish the rest of the body. By nature we too must nourish and nurture ourselves first in order to sustain the resources with which we can nurture others. Communication, nourishment, the power to heal, and the wisdom of right relationships are functions associated with the heart. Life can no longer be sustained when the heart stops beating. In terms of the issues and qualities with which the heart is associated, when these are diminished or cease, the richness and meaning of life is lost and emotionally we can no longer live fully. Speaking metaphorically, we die to ourselves and may feel that we are merely surviving our life. Expression of the feelings of the heart is the essence of humanity. Bob Moore, a renowned healer, once said that expression is the best form of protection; I feel this is particularly true of the heart.

Organs of the Vocal System

The vocal organs consist of the larynx (which is continuous with the tubes of the pharynx above and the trachea below it), five specialized cartilages, many small muscles and ligaments, and the vocal cords or vocal diaphragm. Vocalized sound is produced when the two vocal cords are stretched and set into vibration by the air passing between them. The complexity of the human vocal apparatus and related speech areas of the brain has enabled in humankind the development of this sophisticated, symbolic communication and the modes of expression engendered by it: writing, reading, and creative thought. Our evolutionary progress is closely linked to the development of speech, as are all areas of artistic and creative expression that are uniquely human. The vocal center supports and reflects human creativity and the power to express the truth of who we are.

(The brain and special sense organs are also part of the nervous system, which we will be looking at in the next chapter. However, they are also organs and can therefore be explored in the ways described here.)

Exploration

The idea of bringing awareness consciously into the organs and moving from there is an unfamiliar one to most of us, yet many people find the experience an unusually rewarding one. A whole new world of perception and expression may be opened up, bringing endless opportunities for creative and healing explorations. Learning to access and work with oneself at this level of physiological and psychological process can be very empowering when these deep resources of energy are contacted. We also validate ourselves, our own wisdom and authority, as we develop the ability to facilitate our own process of healing.

Contacting and moving from the organs can give support, energy, power, feeling, and presence to posture, movement, and vocal expression. Holding patterns in the body and mind can be gently released, allowing fluidity and expansiveness to return to our movement. The process can deepen and enrich our experience of ourselves and of life and bring new insight to our quest for self-knowledge. Here are some ways to begin to explore the presence of the organs of your own body in movement.

1. Using pictures in anatomy books to familiarize yourself with their details, locate specific organs in your own body. Visualize them as clearly as possible and explore breathing, hissing, and sounding into each one individually. This helps to stimulate sensation and awareness of their actual presence and location and to increase their tone and vitality.

2. This can also be done with a partner. Focus together on a specific organ in one person (the recipient). Her partner places her two hands over the location of the organ on opposite sides (e.g. front and back of the body), while the recipient visualizes, breathes, etc., as above. The holding should have a full, weighted but sensitive and fluid "organ" quality, as if holding a balloon filled with water; think of holding from your own organs. A partner's presence helps to keep the attention focused and will further concentrate the energy where directed. (Fig. 8.10)

3. Rotate the individual organs in the three basic planes: sagittal, vertical, and horizontal. Explore this in different positions: sitting; lying on your front, back, and side; on hands and knees; standing; or upside down in a shoulder stand.

Let these subtle movements initiate

Figure 8.10 "Holding" the organ to facilitate the release of weight and initiation of movement through them.

whole body activity in rolling, turning, tilting, etc. (Fig. 8.11)

4. While lying on the floor roll slowly from the back to the sides and front of the body, feeling the shifting of the weight to the underside of the organs as you move. Then work with chang-

Figure 8.11
The organs support
movement through the
whole body.

ing levels in relation to the floor, through lying, sitting, squatting, hands and knees, standing, etc. Explore the dynamics of gravity and levity, support, and mobility acting on the organs as the organs initiate and support these changes.

5. Any of the above explorations can be followed by improvised dance movement, which helps to release the mind of "sensing" and enables you to feel and express the full energy and power of the organs. Allow this energy to move you freely and fully through the space: to and from the floor, turning, rolling, jumping, running, balancing, and so on. Follow the organs' own momentum, rhythm, direction, and quality of movement, and let yourself be surprised.

As you finish, you might like to make a drawing; do this spontaneously from your organs and from the energy of the dance. This helps to further integrate and ground the experience. Or you might like to draw your inner impression of all of the organs after you have worked with them. Do this not in a literal way but rather try to capture the feeling of energy, the movement, the colors, the energetic quality and shape, or a personal image of each organ. This can reveal much about the unconscious and psychological messages that the organs may be carrying.

The Endocrine System

The glands of the endocrine system have a profound effect on both physiological functioning and feeling states; they also affect the quality of movement support and expression. The glands are generally much smaller than the organs and form an energetically connected network that lies along the length of and more or less anterior to the spine from head to tail. Energetically they connect to specific bones and joints and give support to the spine at related areas; like the organs, they should give support from within through their fully breathing, open presence and aliveness. Each gland also has a governing effect over particular organs, senses, and perceptions, and expresses a unique quality of mind and feeling, as do the individual organs.

The endocrine glands have been described as being one of the subtlest manifestations of energy in the physical body, relating closely to the *chakras*.[14] They could be seen to act as a link between the subtle and invisible energy body and the manifest physical body. Their energy is less dense than that of the organs, and this enables them to be recognized as a distinct system through sensitive touch and inner listening or through observation of their expression in posture and movement. With a little practice it is not difficult to recognize the difference between an endocrine gland and an organ.

In her classification of endocrine glands, Bonnie Bainbridge Cohen includes several structures that have not generally been recognized as glands. Their activities include the regulation of respiration and circulation and the control of certain mineral levels in the blood; the functions of some are as yet unknown. These structures are termed "bodies" to distinguish them from structures formally recognized as glands. They are included in the endocrine system here because they are felt to express the quality and higher vibration of energy experienced in the glands themselves. Together these glands and bodies form an integrated system of energy that channels the flow of movement along the length of the spine with a particular quality of clarity and alertness.

The cells of the endocrine glands secrete hormones into the bloodstream; these hormones are chemical agents which affect the cells of organs, tissues, or other glands through either stimulation or inhibition of their functions. This change in activity is registered by the brain, which then sends out further messages to regulate the secretion of more or less hormones. Through a highly complex and finely attuned interaction, the endocrine and nervous systems regulate and integrate the functioning of the body at a cellular level. The endocrine system is chemical in nature and older in terms of evolutionary development; the nervous system is primarily electrical, with chemical processes also playing a part, and more specialized. Through their communication and responses they control growth, reproduction, and metabolism, and also affect the mind states associated with these processes.

The way of working with the glands is similar to that with the organs; first pictures are studied where available (some of the "bodies" may not be found in anatomy books), and the glands are located in the body. They are highly sensitive; some people may find that simply placing their attention on the glands will be enough to awaken awareness there and stimulate their energy. A light fingertip touch is most helpful in locating the glands, as this enables us to feel with precision and sensitivity. Breathing, hissing, and sounding into the glands are ways to further deepen the contact. Each has its own vibration, tone, and rhythm; you can explore to find which sounds and rhythms most clearly resonate with the energy of each gland. It is important when working with the glands not to overstimulate one in isolation from the rest of the system; they are extremely sensitive and powerful, and overuse or overstimulation can create imbalances. For each person the length of time spent working with one gland in these ways may vary, depending on one's sensitivity and ability to make contact with this system.

As with the organs, movement can be initiated in the glands first by finding the subtle shifts and rotations that happen internally as the mind images and directs the action. These small movements can free any holding that is taking place within or around a gland, and can realign the gland with the endocrine system as a whole. This also affects the alignment and mobility of the organs, spine, and related joints, and influences the patterning of muscles and ligaments by releasing, connecting, strengthening, or increasing their range of movement. The endocrine system is central to the functioning of the body-mind on all levels, and even the subtlest change in their alignment and openness can have a profound effect on posture and movement, and on states of feeling, perception, and awareness.

Small movements initiated in the glands can then be carried into movement through the whole body; this is an important step in the process of repatterning. As the energy of the glands is awakened and freed it must then be channeled through the whole body to be integrated and useful. Without this integration, stimulation of the glands can cause the chaotic aspect of their feeling nature to be evoked. The most direct way to integrate this energy, and at the same time more firmly establish the new movement patterning, is through exercises that encourage a sequential flow of energy along the spine and through the limbs. The movement, felt to be initiated or supported by one gland primarily, travels sequentially through all of the glands and along the whole length of the spine or flows out through the arms and legs. We are following the natural courses of movement in the body; it is helpful not to let the mind get too focused on an area of blockage or weakness, as this can tend to feed the problem. Instead, let the energy flow and recreate its own natural channels.

If the movement does not flow through a certain area of the body and there is a sense of solidity or immobility there, we can work with the glands relating to that part. Then we gradually free a channel through the endocrine system as a whole. Special attention is given to opening the glands of the tail and head so that energy is not caught within the system but can extend and express freely through movement in space, in interaction with both earth and heaven. Awareness of the hands and feet as exit and entry places for this flow of movement also helps to both free the flow of energy within and relate the inner movement to the outside environment. The energy of the glands is grounded and contained by being expressed through movement in this way.

Relationship to the Developmental Patterns

It was mentioned earlier that the glands have a special relationship to the Developmental Movement patterns. Each gland gives energetic support to one of the patterns in particular, although ideally the whole system should be active as the flow of movement travels along the length of the spine. The glands give to the patterns a quality of lightness and ease that is simultaneously fiery and alert; they create clarity of form and flow in space. They tend to express the *crystallization* of energy in clear movement forms; the organs, in comparison, are more about inner process, emotional feeling, and sensation, and their form in movement is less clearly defined or crystallized.

In the early stages of developing a new movement pattern, the infant is very much involved in the inner, organic processes

of sensation. Weight, gravity, touch, sensory, kinesthetic, and proprioceptive sensations are its concerns. The organ "mind" of self-
absorption and self-reflection is present, along with the emotional
feelings of challenge or frustration inherent in learning something new. The final mastery and crystallization of a pattern, the
breakthrough to a new level of functioning, requires the activation and support of the endocrine glands. With their support, the
infant transitions from self-awareness and inner focus to a more
alert and spacious perception of the external environment. The
glands bring in curiosity and openness towards the external world.

In each pattern a particular gland, or pair of glands, supports
the spine at the place where there is maximum stress, likelihood
of disconnection, or tendency to collapse in that movement and
posture. The powerful energy of the glands exerts a pull against
gravity; this is focused in the area of the spine in front of which
the gland lies, where the stress is greatest. But it is when the energy
moves *through* the gland and is channeled through the whole system of glands sequentially that support is most effective. We might
imagine each gland to be a gateway that allows or inhibits the
passage of energy through it. This *throughness* of the flow of energy
is essential to integrating the whole body into the movement pattern; it is the coordination of the endocrine system with the neuromuscular system in the Developmental Patterns that provides
the basis for strength, clarity, and graceful action. The glands
can stimulate and support the movement; the movement itself
also stimulates the glands. So in working to integrate the systems,
we can use either the outer form of the movement or the inner
sensation of the glands' energy as our starting place. From either
approach we work toward the alignment of movement and feeling.

The Creative "Mind" of the Endocrine System

The endocrine system concerns intuition, feeling, and inner balance or chaos. It touches deeply into the core of who we are and how we perceive and express ourselves in the world. Issues relating to personal, social, creative, and spiritual dimensions are expressed through the different glands.

Working with the glands through movement expression, either in specific exercises or free dance exploration, helps to balance the endocrine system as a whole. Bringing the feeling content of their energy into expression through creative activities, such as dance, music, theater, drawing, writing, or ritual, helps to bring into awareness and integrate these deep impulses. The endocrine glands are a source of great creative energy and contact with them enables us to access the movements and imagination of the hidden layers of ourselves. While each gland expresses qualities of "mind," feeling, and movement that seem archetypal and universal, different individuals or cultures will discover unique images and forms through which to convey these qualities. This tendency of both the deeply personal and the universal means that the endocrine system is an important vehicle for communicating our innermost personal intuitions and experiences in creative ways that can be recognized and shared by others.

In the process of artistic creation we might move between the organ and endocrine systems; in making a dance, for example, the early stages may involve much inner reflection and processing of emotional material or ideas into movement—the work is often described as evolving "organically." When the dance is finally made and performed before an audience, we see the expression of these feelings, images, and ideas within a clear crystallization of formed movement. The tension of the performance situation also stimulates a greater level of alertness, awareness, and expressiveness associated with the endocrine system. Of course, each individual's pattern in regard to this is different and some performers express more powerfully through the organs or some other system. The audience, too, may be touched through any system, depending on the variables of their own and the performer's tendencies. Nonetheless, even if they are unconscious to it, their bodies will resonate through the systems, glands, or organs being most clearly expressed by the performer. If a musician plays from his or her heart, it is in our own heart that we will quite literally be moved, experiencing both a kinesthetic and an emotional sensation.

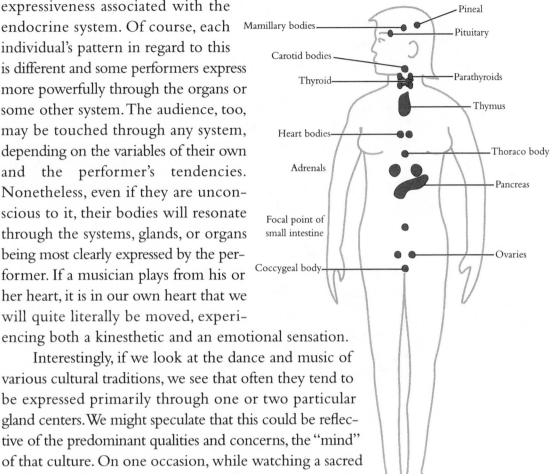

Mamillary bodies
Pineal
Pituitary
Carotid bodies
Thyroid
Parathyroids
Thymus
Heart bodies
Thoraco body
Adrenals
Pancreas
Focal point of small intestine
Ovaries
Coccygeal body

The endocrine glands and bodies.

Interestingly, if we look at the dance and music of various cultural traditions, we see that often they tend to be expressed primarily through one or two particular gland centers. We might speculate that this could be reflective of the predominant qualities and concerns, the "mind" of that culture. On one occasion, while watching a sacred Tibetan Buddhist lama dance, I was fascinated to see that every center of the dancers' bodies was present and involved in a harmonious balance, an expression in movement ritual of the inherent wholeness of the dancers' body-mind.

The Glands:
Structure, Location, and Function

The following descriptions of associations are adapted from research and writing by Bonnie Bainbridge Cohen.[15]

The **coccygeal body**[16] is a small, irregularly shaped cluster of cells lying at the tip of the coccyx (tailbone). It is richly supplied with blood vessels and nerves, making it an area of potential vitality and sensitivity. The function of the coccygeal body is scientifically unknown, but when energetically active it supports the pelvic floor and frees the tail of the spine and the hip joints, giving lightness and agility to movement of the legs. When the coccygeal body is active and the hip joints free, a lively connection will be felt in the toes. The coccygeal body supports the connection of the legs into the spine in the Homolateral Push patterns from the feet. It connects us to our roots within the earth, giving a groundedness that is based on self-love and the will toward personal survival, and that underlies instinctual love. The "mind" is of nondifferentiation between inner and outer worlds, closely associated with fetal life and the trance-like state of unity experienced through group ritual. It relates to the root *chakra* and serves as a grounding force for states of higher spiritual unity associated with the glands of the head. Energetically the coccygeal body grounds the head glands.

The **gonads** are the sex glands of both males and females. In girls and women there are two ovaries, each about the size and shape of an almond, lying in the center of the pelvic cavity front to back, midway between the navel and pubic bone, and about two to three inches out from the midline on each side. They secrete the female sex hormones that regulate the development of female sex characteristics, the menstrual cycle, pregnancy, labor and childbirth, and lactation. As organs, they also produce the ova, or egg cells.

In boys and men the two testes, about one and one-half inches long and about half as wide, lie in the scrotum external to the pelvic cavity. In working with these glands as support for movement we use approximately the same location as in women, at two points on the vas deferens (the tube which connects the testes to the penis) deep in the pelvic cavity. This is close to a location within the body through which they pass as they descend to their external position. The testes secrete the male sex hormones that regulate male sex characteristics and reproductive functions; they also produce and store sperm cells.

The gonads underlie the instinct for physical creativity, unity, and the expression of shared sensuality, sexuality, and physical love. They bring us to a sense of home within our bodies and on the earth. Energetically they connect to the heels, ankles, forelegs, and sacrum; the alignment of these parts with the gonads gives grounding and support for the pelvis and lower body. The gonads support the Homolateral Push from the hands by directing the energy from the hand through the pelvis and down to the foot on the same side. The gonads also ground the glands of the throat and are closely associated with them. Both are centers of creativity: the gonads in relation to the earth, sexual expression, and procreation; the throat glands relate with verbal, vocal, and artistic expression. The gonads are related to the sacral *chakra*.

The **small intestine** is not a gland as such, but within the lining of the digestive tract are scattered cells that secrete hormones.[17] Evidence as to their specific function is not yet certain. Their inclusion within the endocrine system gives support to the Mouthing Pattern of early infancy, the organs of the digestive tract, the lumbosacral joint, and the curve of the lumbar spine. A point in the center of the intestinal area just below the navel and corresponding to the *hara* (a Japanese term for the body's energetic belly-center), as well as the center of weight and gravity, can be used as a focal point of location. This "gland" underlies all aspects of nourishment and self-support.

The **adrenals**, two glands one to two inches long, sit on top of each kidney. They lie on either side of the spine at the level of the eleventh and twelfth ribs beneath the diaphragm and in front of the strong muscles of the back wall. The medulla, or inner portion, secretes adrenaline, the "fight or flight" hormone that prepares the body to deal with life-threatening situations, and norepinephrine, which counterbalances this activity when the stressful situation has been resolved. The cortex, or outer portion, secretes hormones called steroids that regulate mineral and blood sugar levels. The adrenals are associated with the life force, the instinct for survival and acts of instinctive physical courage. Fear (as a protective function), rage, and anxiety are also experienced here. Stress and exhaustion can build up in the adrenals if their energy is continually overused or is not released in appropriate activity. They ground, support, and are supported by the glands of the chest area and connect energetically to the knees, femur, and sacroiliac joints. Their energy can also be felt to radiate throughout the whole body as a support for the Navel Radiation pattern.

The **pancreas** is about six inches long and shaped like a fish. Its head lies in the center of the torso at about the level of the second lumbar vertebra, the body points up and back to the left, and the tail touches the spleen close to the back wall of the rib cage. It functions both as an organ and an exocrine gland as well as an endocrine gland. The endocrine cells are mainly situated near the head of the pancreas, which lies midway between the navel and the lower tip of the sternum, or breastbone. This location is at the center of two triangles formed by connections between the head and feet and the tail and hands; the pancreas gives energetic support to these six extremities, and maintains spatial tension between them. In this way it also supports the Homologous Push from the feet, where for the first time all of the limbs come into full extension and spatial tension throughout the whole body is established. The pancreas secretes insulin and glucagon, which lower and raise the blood sugar level.

The energy of the pancreas, as an organ and gland of the solar plexus *chakra,* can be felt to underlie the social instinct; its place below the heart offers a base of support for the transition from self-concern and awareness *of* the group to concern and love *for* the group, which happens as consciousness moves upward from the solar plexus to the heart *chakra*. The solar plexus and the energy of its organs and glands are concerned with personal power, personal needs, and the survival of ego in relation to the other and to the collective; these are the necessary bases for an emergence into heart awareness. If issues pertaining to the solar plexus are confused with those of the heart or sacral centers, then love and sexuality will be expressed through competitiveness, conflict, and neediness. The pancreas can express strong emotional energies, such as anger, fear, excitement, and exuberance, and in its endocrine function it energetically governs the organs of the solar plexus area.

The **thoraco body**[18] lies close to the diaphragm, about three-quarters of an inch behind the xiphoid process, the tip of the sternum. Its openness supports the thoracic diaphragm and external respiration through the lungs. The thoraco body is a gateway between the upper and lower centers of the body; it plays an important role in the relationship and integration of the solar plexus and heart centers and the transition from self- to group-consciousness. It supports the fullness of breath throughout the body, and governs the Pre-Spinal pattern. The thoracic body thus relates to the alignment and integration of the whole physical body.

Like the thoraco body, the **heart bodies** were first discovered experientially. They are felt to lie behind the sternum, one on each side, within or on the heart organ at the level of the nipples or fifth ribs. Recent research is showing that the heart itself has an endocrine function. The heart muscle has been found to secrete hormones that affect circulatory homeostasis, among other functions.[19] The felt presence and function of two heart bodies as discrete structures attached to or lying within the heart might be related to this, but more research would be needed before this could be confirmed. Energetically the heart bodies support gestures of opening and embracing with the arms, as well as expression of feelings of the heart in reaching out through the hands and eyes. The heart bodies also support the forearms and wrists and govern the Homologous Push from the hands. They underlie the feeling and expression of love for others and group consciousness. Like the heart organ, they are associated with giving and receiving, unconditional love and acceptance, and union with another. The energy of the heart bodies, when unexpressed or frustrated, can turn back in and down to manifest in strong emotional forms, such as depression, hurt, or anger, through the solar plexus center.

The **thymus** is a two-lobed gland lying directly behind the manubrium, the top part of the sternum, and above the heart. It is shaped rather like a butterfly, and is usually about two inches in length and width in adults, although sometimes it may atrophy to no more than a small patch of fibrous tissue. In childhood it is very much larger but shrinks at puberty; its cells migrate to other parts of the body where they assist in the immune function of the lymphatic system. It may also shrink or atrophy through lack of proper stimulation and use, both energetically as a support for posture and movement, and physiologically as the body's natural means of defense. The thymus plays an important role in the body's mechanism of defense against disease, producing T-lymphocytes and a hormone which stimulates their development. As part of the lymphatic system the thymus energetically helps create the sense of personal boundaries. The experience of loss of boundaries and protection often accompanies a breakdown of the immune system; stimulation of the thymus gland can help to strengthen the weakened boundaries, and facilitate the transformation of fearful feelings into courageous action. It has a close relationship to the adrenals, the center of instinctive courage. The thymus particularly gives support and freedom of movement to the shoulder joint and shoulder girdle, as well as a feeling of width and openness across the front of the chest between the shoulders. When actively supporting it takes us forward and heavenward. This open posture itself expresses a courageous state of mind. The thymus supports the Homologous Reach and Pull from the hands. Lying just above the heart *chakra,* it underlies the expression of courage based on love, beyond instinct or duty.

The **thyroid** is a large fleshy gland that wraps around the middle and lower portions of the front of the throat. Its two lobes, each about two inches in length, lie to each side, and they are connected by a narrow band across the front of the trachea just below the thyroid cartilage. The thyroid gland secretes hormones that play an important role in regulating metabolism and also mental and sexual development. This gland supports the power of the voice in singing and is related to the throat *chakra* which is the center of artistic and creative expression. In this, and in its physiological effect upon sexual development, it has a special relationship to the gonads, the center of physical creativity. Energetically the thyroid also gives support to the elbows and the humerus; moving the thyroid and the elbow joints counter to one another can greatly open and free this area. We see this expressed beautifully in Indian and Balinese classical dance, where this fine articulation is also grounded by a strong focus on movement through the pelvis and heels (gonads). This gland also supports the Homologous Reach and Pull from the feet.

The **parathyroids** are four and sometimes more small oval discs, each about one-third of an inch in length, embedded in the back of the thyroid. There are two inferior and two superior parathyroids. They control the calcium level in the blood. Like the thyroid, they underlie creative expression, giving support to the gentle quality of the singing voice and the refinement of its articulation in melody. They also help to integrate vocal expression with finely articulated movements of the hands, again seen in Asian classical dance forms. Movement of the ribs and between the ribs and scapulae is supported and energized by the parathyroids; the freeing of the scapulae from the ribs is a necessary support for the Contralateral Reach and Pull patterns.

The **carotid bodies** are shaped like two small beans and lie within the bifurcation of the carotid artery, on each side of the upper neck just under the angles of the jaw. They contain nerve endings that respond to pressure and chemical changes in the blood, thus playing an important role in the regulation of respiration and circulation. The carotid bodies do not have any recognized endocrine function, but energetically they provide an important gateway between the glands beneath and above them. The free flow of energy through them helps to balance and integrate the head with the rest of the body. They give support to the neck, the vertebral column, and the Spinal Push pattern from the tail. They also give power to the speaking voice through support of the silence surrounding the sound and they underlie a sense of "divine nobility" and the courage to express one's truth.

The **pituitary** is a single gland about the size of a small pea, with an anterior and posterior lobe. It hangs by a stalk from the third ventricle of the brain and lies beneath the cerebral cortex within a small depression in the sphenoid bone, just behind the top of the nasal cavity and slightly in front of the midline of the head. It used to be known as the "master gland," as it was thought to direct the entire endocrine system. This system is now coming to be understood not as hierarchical but rather inter- and intracommunicative in nature, with each gland constantly responding to and counterbalancing activities elsewhere in the neuroendocrine system. The pituitary affects the secretion of hormones by the thyroid, adrenals, and gonads, and it also secretes hormones that induce labor, lactation, cell growth and division, and that influence metabolism and the retention of water by the kidneys.

The pituitary supports the functions of the eyes and all processes related to vision, intelligence, and imaginative and conceptual thought. Overuse of this gland tends to pull the head slightly forwards of the midline, in the posture typical of one concentrated on visual and mental tasks such as reading and writing, activities associated with the pituitary. This pulling forward of the head can be countered by anchoring the pituitary in the tail of the spine; the pituitary can thus be felt to support the Spinal Reach and Pull from the tail. This gland is also considered to underlie altruistic love and compassion.

The **mamillary bodies** are two small, round structures lying side-by-side within the midbrain area, slightly behind and above the pituitary and aligned with the midline of the body. They have not been traditionally recognized as endocrine glands, although recent research has been carried out into their possible endocrine function. They are part of the limbic system, an area of the brain that secretes chemical substances into the bloodstream in a way similar to the endocrine glands proper; the pain/pleasure response is connected with this function. Because of their close relationship to both nervous and endocrine activities, the mamillary bodies can be considered to be a point of clear access from one system to the other. We might also call them the "keystone" of the neuroendocrine system. They govern the mouth and nose, and are associated with the primitive functions of the sense of smell and the sucking and swallowing reflexes. They are also associated with the activation of alertness: their energy initiates the first Spinal Reach and Pull pattern through the head, which awakens us to higher levels of attention and perception.

The mamillary bodies underlie insight, perception, and the expansion or dissolution of the boundaries of time and space. Centering awareness in the mamillary bodies helps to align the head and body along this vertical axis, and allow the top of the head to reach into the space above. This supports balance in off-balance movements, such as turning, rolling, and spiraling in space. Such centering can also produce pleasurable feelings of spaciousness, timelessness, and an openness of the sense perceptions similar to those experienced in some types of meditative practices.

The **pineal**, the last and first of the glands, is a small oval or cone-shaped structure lying on a diagonal up and back from the pituitary and mamillary bodies, above the midbrain. It secretes a hormone called melatonin; its secretions occur primarily at night and are inhibited by light. The pineal, via melatonin, is beginning to be viewed as a "central regulator" of neural immune, endocrine, sexual, thermoregulatory, and other body functions.[20] It plays a strong role in harmonizing patterns of rest and activity, menstrual cycles, sleep, etc., with the changes of the sun and moon in daily, monthly, and yearly cycles. Sensitive to vibration, it governs the functions of the ears and is associated with both the sense of hearing and the vestibular, or balance, mechanisms of the inner ear. It is active in preventing premature aging; in fact, according to researchers in Italy and Russia, there is now "proof that aging *initiates and progresses* in the pineal gland itself."[21] The pineal also appears to be strongly involved in Down's syndrome.[22] Clear functional links have been established between the pineal and other glands such as the thymus, thyroid, and pituitary, particularly in relation to immune function.[23]

The pineal is also traditionally thought to be the mystical "third eye," the eye that looks within, and does in fact contain cells similar to those found in the retina of the eye itself. Alan Bleakley writes, "There is evidence that the pineal gland does 'see' things from inside the body . . . [and] translates bodily phenomena into internal image, as in dreams or visions."[24] Its development is encouraged in those training in healing practices and those involved in spiritual disciplines, and has been reported to be well-developed in people with clairvoyant or psychic abilities. The pineal is both the birthing gland and the place of exit into death. It supports the Spinal Push from the head, which initiates the birthing pattern. It is also considered to underlie the transcendence of duality and separateness in spiritual practice. Centering in the pineal gland can evoke a sense of the depth of time, of ancient history and eternity, past and future brought together in the present.

Activating the glands of the head to initiate and support

movement gives a sense of fine articulation to the balancing and moving of the head; this allows it to move as part of the body as a whole, which releases energy in the spine and gives to the movement a more fluid, dynamic, and integrated quality. When all of the glands are active and their energy is in balance, they can be aligned with each other through subtle inner adjustments. The spine and other structures of the body will then automatically fall into a clearer alignment around this integrated core. When working with the endocrine system, it is important to remember to open up the channels among all the glands and bodies so that the energy released is able to move through and express; working with individual glands in isolation from the whole is not usually helpful. Movement and the use of imagery can facilitate this release of energy through the body, and help integrate changes taking place. Balance of this system supports integration and harmony within the person as a whole, and a sense of the depth and fullness of being.

Exploration

Conscious use of the endocrine system can help to create a clearer and more integrated alignment of the physical body and of the body with the mind and feelings that are expressed through it. The glands have a powerful effect on other body systems, profoundly affecting skeletal, muscular, and organic patterns of both movement and support. Greater openness, flexibility, strength, and fluidity in these systems can be facilitated by repatterning through the endocrine system. Alertness, clarity, and vibrancy are felt when this system is actively supporting and expressing in movement. It can also open us to a wealth of feelings, images, symbols, memories, and dreams that inhabit the inner world of our psyche and are a source of creative and transformative work. Deepening our awareness to the endocrine system also gives us access to our intuition and the source of our inner wisdom.

1. As with the organs, locate the individual glands in your own body, or, if possible, do this with a friend's help. You can use the techniques of visualization, breathing, hissing, or sounding into each gland—explore the different sounds and rhythms that each gland evokes. Work gently; the glands are highly sensitive energetic structures and it is important not to force them. Listen to your individual needs in relation to them.

2. Initiate small movements, shifts, and rotations in different directions through the glands. Then take these initiations into a fuller range of movement involving the rest of the body more actively. To do this you will have to bring in other systems, such as the muscles and blood; through them the energy of the glands is expressed and made visible in whole body movement.

3. Practice the Developmental Patterns with the support of the appropriate glands. (See Table 1, pages 84–85.)

4. It can be exciting to experiment with dancing from the glands, being aware of the movement qualities, feelings, images, perceptual states, etc. that each one evokes. You might also explore dancing to different kinds of music and seeing if you can identify from which glands your movement is coming.

5. The following sequence of exercises includes some of the Developmental Patterns. Practice each one slowly and attentively a few times, with your focus on the gland or sequence of glands that support and initiate the movement, as described. Then do the whole sequence in a fluid and continuous way without breaks, and try to feel the flow of energy through the glands, spine, and body as a whole. Each movement can be done only once or repeated as many times as you like. Do this with a feeling of ease and a very light quality of attention to the glands as the movement passes through them. Here you are looking for the flow and connectedness of the energy through the glands, and also through the spine. Once the exercises are familiar, allow yourself to feel the flow of the movement from within rather than being too attached to the outer forms.

You may also like to try this sequence initiating from the organs, bones, or muscles; notice any changes in quality.

Feel free to adapt or add to this sequence according to your body's own needs. After working with the exercises for a while in a focused way, you may gradually begin to improvise using the sequence as a framework from which to explore your own creative movements.

Movement Sequence to Activate and Balance the Endocrine System

1. Lying on your back with knees bent and feet flat on the floor, initiate a rotational movement in the coccygeal body which swings the tail upwards in a small arc toward the pubic bone then downward to the floor. Gently press the toes and balls of the feet into the ground as the tail lifts up, keeping the abdominal and thigh muscles relaxed; the lower back should feel lengthened and pressed closer to the floor. The pressure through the feet supports the tail as it lifts. Repeat this movement several times. (Fig. 8.12)

2. Initiating in the same way with the coccygeal body, develop this movement by reaching the tail up and out toward the knees; then rotate and reach through the gonads, small intestine center, and adrenals sequentially, so that the pelvis tilts up and is then lifted off the floor, arcing the spine into a "bridge" position. Move as you exhale, breathing the energy down through the toes as the tail lifts, through the heels as the gonads and pelvis rotate and lift, and through the arches of the feet as the small intestine initiates, giving greater support; the energy from the adrenals radiates out through the

Figure 8.12
Initiate through the coccygeal body.

225

Figure 8.13
Coccygeal body, gonads,
small intestine and
adrenals.

Figure 8.14
Pancreas through to
glands of the head.

Figure 8.15   Pancreas to the hands and feet.

knees, which lengthen away. Come down again by reversing the order of initiation so that the spine lengthens sequentially back onto the floor. (Fig. 8.13) Repeat.

3. (The following two exercises can be done at the end of the sequence if preferred.) From this "bridge" position, breathe into the pancreas and as you exhale, send the energy up through all the glands into the head and hands; with this impulse push yourself up to balance on your hands and the top of your head. (Fig. 8.14) If your back is limber and healthy, on the next exhalation send the energy out again from the pancreas through the hands and feet, and push up through the hands and feet into a high back arch. If this strains your lower back, omit this part of the exercise until greater flexibility is available. The pancreas is the "keystone," or highest point of the arch, and the hands and feet continue to support the pancreas by pressing into the floor. In this position think of lengthening the spine from both ends by *taking out the slack,* or rocking gently head- and tailward. (Fig. 8.15) Then slowly lower back to the floor from the head down to the pelvis and tail, lengthening through each gland in turn. Rest for a moment before repeating or going on to the next exercise.

4. Using the support of your hands if needed, reach the tail up and back over your head so that your feet

Figure 8.16   Adrenals.

Figure 8.17   Gonads.

touch the ground behind, in the yoga "plough" position. Then initiate with a rotation of the adrenals to bring the middle area of the spine upright (Fig. 8.16), a rotation of the gonads to align the pelvis over the spine and shoulders (Fig. 8.17), and a reach of the coccygeal body forward and up toward the toes to lengthen the legs into a full shoulder stand. (Fig. 8.18) Reverse the rotations to curve the feet and knees back to the floor behind your head. Then from the pancreas reach through the legs to extend them fully out along the floor behind you. (Fig. 8.19) Release the pancreas and repeat the shoulder stand, initiating again with the adrenals, gonads, and coccygeal body. After a few such movements, unroll the whole spine back onto the floor and rest.

    5. Rotate and reach the coccygeal body toward the right side, simultaneously extending through the toes of the left foot so that it reaches over the right leg toward the floor on the opposite side. Rotate each gland in the same direction, in sequence one by one from the coccygeal body up to the glands of the head, to take you onto a spirallic roll onto your front. The upper body, shoulder, arm, neck, and head should remain relaxed and be pulled through a diagonal stretch with each gland in turn supporting and initiating the movement of the spine. (Fig. 8.20) To return, initiate

Figure 8.18   Coccygeal body.

227

with a rotation of the pineal and then move down sequentially, rotating and reaching through each gland in turn. This time, the roll begins with the turning of the head and reach of the left hand to the floor behind. Repeat on the other side and alternate sides several times.

6. Roll onto your front and lie with the arms loosely on the floor above your head. Breathe into the thoraco body and feel the breath expand the whole rib cage, and move up into the arms and head. Exhale and slide the arms gently along the floor close to the head, until they are extended. Then, as you begin to inhale, slightly lift the upper chest, head, and arms off the floor. There should be no straining to lift high; let the fullness of the breath create an easy and integrated movement, the arms and head carried by the rib cage and spine. Then pull the arms out wide to the sides, a little raised from the floor, as you complete the inhalation. (Fig. 8.21) Pause between the inhalation and exhalation as you relax gently to the floor again. Exhale and repeat the exercise. As you rise up you should feel that the thoraco body and lower rib cage are grounded and act as a base of support for the upper body by remaining in contact with the ground; this avoids strain in the lower back and tightening of the buttock muscles. The legs should rest easily on the floor throughout. This movement is like swimming the breaststroke.

7. Beginning with a lengthening through the top of the head initiated in the pineal, sequence down through each gland to the heart bodies, rotating and reaching up through each in turn to lift the head high on the support of the forearms. (Fig. 8.22) Let the arm and shoulder muscles do as little work as possible; feel the spine lengthening up between the shoulder blades and the spine, rib cage, scapulae, elbows, forearms, and wrists support in turn, as the initiation sequences down through the glands from pineal to the heart bodies. Come to the floor again by rotating and reaching through each gland from the heart bodies to the pineal, then repeat.

8. From the "sphinx" posture, push back into the prepara-

Figure 8.19   From this position the pancreas extends the legs out along the floor.

Figure 8.20   All glands in sequence from coccygeal body to pineal, and pineal to coccygeal body.

*Figure 8.21
Thoraco body.*

*Figure 8.22
Pineal to heart bodies.*

229

Figure 8.23
All glands sequentially, from coccygeal body to pineal.

8. From the "sphinx" posture, push back into the preparatory position for the Spinal Push patterns. Initiate a rocking forward and back through the spine by pushing from the tail, then head (see the description of the Spinal Push patterns in Chapter Three). Let the energy flow through all the glands, from coccygeal body to pineal, as you push forward onto the top of the head (Fig. 8.23); then from pineal to coccygeal body as you push back onto the heels again. (Fig. 8.24) The carotid bodies give additional support to the cervical spine in the push from the tail; the pineal grounds and supports throughout the whole movement sequence in the push from the head.

9. Reach the head and spine forward along the floor and up through the mamillary bodies onto the support of the hands, as in the Spinal Reach and Pull from the head (Fig. 8.25); then pull back onto the hands and knees, reaching through the coccygeal body. (Fig. 8.26) The pituitary supports the head in the movement backward. Pull all the way back into the starting position and repeat. Then, on hands and knees, rock forward and back by pulling through the head and tail, feeling the energy flowing sequentially through all the glands so that the spine is moving not as a rigid unit but lengthens and has fluidity.

Figure 8.24
All glands sequentially, from pineal to coccygeal body.

10. Through the coccygeal body, pull the tail up towards the ceiling so that you come onto your hands and feet in a "triangular" posture. Push forwards from the feet to the head so that the body lengthens forwards, with weight on the hands,

Figure 8.25
Mamillary bodies.

Figure 8.26
Coccygeal body initiates,
pituitary supports.

Figure 8.27   Pancreas.

Figure 8.28   Pancreas.

and is supported in an integrated line from feet to head. (Fig. 8.27) Then go back into the triangle position by pushing from the hands into the tail. (Fig. 8.28) The pancreas supports both movements by maintaining spatial tension between the extremities, so that the spine does not collapse in the middle. This movement is quite vigorous, and the spatial tension between the six extremities of head, tail, hands, and feet should be maintained throughout; the pancreas lies at the center of the two triangles made by the hands and tail and the feet and head.

11. Walk the hands back toward the feet and relax the arms and shoulders; then unroll the spine into standing. Begin with a rotation of the coccygeal body to bring it under the pelvis; it then supports as the gonads rotate to come into alignment with it, bringing the pelvis into a vertical position. The gonads then support as the small intestine center rotates and moves into alignment above them, and so on up the spine to standing, each gland in turn bringing its related area of the spine into vertical alignment. (Figs. 29–31) The neck is relaxed, bringing the head up last.

12. Keeping the spine soft and fluid, make a small circular movement with the top of the head in the horizontal plane, as if you were drawing a circle in the space above you; this is initiated in the mamillary bodies, and the pivotal point of the movement

Figure 8.29   Coccygeal body to pineal.

Figure 8.30

Figure 8.31

Figure 8.32   Mamillary bodies.

Figure 8.33   Mamillary bodies.

Figure 8.34
Mamillary bodies.

should be in the ankles and feet. The whole body is felt to rotate rather like a gyroscope. (Fig. 8.32) Gradually increase the range of the movement to the edge of your balance (Fig. 8.33), then let your head swing forward and down in a lunging motion (Fig. 8.34); the mamillary bodies continue to initiate the reach of the head through space. Let your hands catch the weight of your falling body; "walk" them forwards with the momentum of the fall as the head continues to reach out, until the whole body is extended. (Fig. 8.35) Repeat steps 11 and 12 if desired.

13. Lower yourself onto the ground, face down, with arms and legs outspread. (Fig. 8.36) Imagine breathing through the navel into the adrenals, and let the movement of the breath radiate out from the center through all six extremities: the head, tail, hands, and feet, as in the Navel Radiation pattern. As you feel the subtle expansion of the breath through the body, extend all of the limbs outward along the floor; when they have lengthened to their full extent, keep reaching

Figure 8.35   Mamillary bodies.

Figure 8.36   Adrenals.

and let the head, tail, and four limbs rise simultaneously off the ground lifting you into the "airplane" posture. (Fig. 8.37) Rise as you inhale and gently lower to the ground as you exhale; repeat several times. Check that there is not too much effort in the lower back muscles or straining in the hips, shoulders, and neck. The spine should feel as if it is lengthening as the energy and breath flow through it out-

Figure 8.37
Adrenals.

ward from the navel. Let the limbs "ride" on the spine and the expansion of breath.

14. Repeat the diagonal rolling exercise (Number 5 above), this time initiating the roll onto the back by rotating and reaching back through the coccygeal body and foot, and the roll onto the front initiating with the hand, eyes, and glands of the head. Before reaching with the hand you can connect the heart bodies, thymus, and thyroid to the hand; initiate a small sliding movement in these glands toward the hand and imagine energy radiating out from them through the

Figure 8.38
Heart bodies, thymus, and thyroid connect to the hand.

arm to the fingertips. The hand will begin to reach out along the floor. Turn to look at your hand and simultaneously initiate the reach of the hand across the body and the rotation of the eyes

235

Figure 8.39
All glands in sequence
from coccygeal body to
pineal and pineal to coc-
cygeal body.

and head, to roll over onto the front. (Fig. 8.38) Sequence through all of the glands as before, from pineal to coccygeal body as you roll onto your front, and coccygeal body to pineal as you roll onto your back again. (Fig. 8.39) Roll to each side alternately.

15. Finish by rolling once again onto your back and rest or, if you wish, to prepare to do the sequence again. (Fig. 8.40)

Figure 8.40
Rest or prepare to begin
again.

Chapter Nine

The Systems of Communication and Transformation

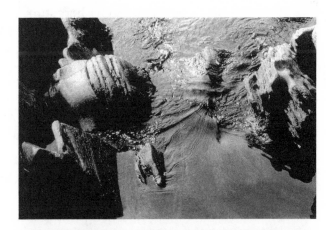

Whether we move or are still, and whatever the quality of that rest or activity, we are expressing or initiating through some aspect of the nervous or fluid systems. They are the means by which the body-mind is organized to function as one individual whole. Through them the other systems are controlled, regulated, nourished, stimulated, communicated with, expressed, and transformed. No aspect of our existence can function without the presence and activity of these two systems.

In the previous chapters we have been exploring the nervous system, either directly or indirectly, in its function of organizing movement patterns through the other body systems and in the unfolding of the developmental process. Wherever a change occurs, whether at a cellular level or in the repatterning of the bones, muscles, or organs, it is mediated through and registered by the nerves and the brain. The registering within the nervous system of new movement sensations means that the activity creating that experience becomes available to the individual as a new choice of action, a means of expression. The nervous system registers all new sensations coming to it and directs responses based on the "memory" and the perception of past experiences. An ongoing process of activity, perception, response, and regulation of further activity is always occurring. This is a highly complex system, with millions of nerve cells involved in any one apparently simple activity. Through the nervous system's sophisticated process of communication, the organization of bodily functions, movement, and perception are coordinated.

The fluids of the body are the medium through which life is nourished and sustained. They are the internal oceans and rivers through which we live; without their constant flow and replenishment, life will cease. In the words of Stanley Keleman, "Living is movement, another word for it is process. Living your dying is the story of the movement of your life."[1]

The fluids concern communication, nourishment, breakdown, renewal, and defense, the process of change and transformation, the living and dying of each moment. They relate to both the survival and the quality of life, and to the balance of rest and activity. The fluids embody the process of expressing wholeness, in whatever form that takes for each individual.

The Nervous System

The nervous system is made up of billions of microscopic nerve cells, each with a cell body and one or more fine projections that vary considerably in length; the longest, the sciatic, runs from the lumbar spine down to the toes. They are bundled together in connective sheaths to form the nerve fibers that are visible to the naked eye in dissection. The projections of the nerve cells conduct impulses to and from the cell bodies by chemical and electrical processes. "Dendrites" current impulses toward the cell body; "axons" transmit messages away from the cell body. Fluid chemical agents called "neurotransmitters" convey impulses from one nerve cell to another over gaps between them called "synapses," passing from the end of the axon of one nerve cell to the beginning of the dendrite of another. In this way messages can be conveyed throughout a complex network to all parts of the body. (Fig. 9.1)

The number of possible connections among the nerve cells of the brain alone is almost infinite. Carl Sagan claims on evidence of a mathematical calculation that the number of synapses between cells in the human brain can potentially give rise to a number of different mental states and functional configurations that is actually greater than the number of elementary particles

Figure 9.1
Conducting nerve cells (neurons) and synapse, showing the direction in which the impulse travels.

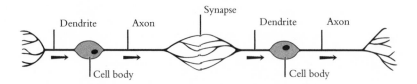

in the universe![2] Whether we accept this statement as fact or only as imagination, it nevertheless conveys a powerful image of the truly awesome potential for choice within the human nervous system and variety in human behavior and experience.

Between the conducting nerve cells, or neurons, of the central nervous system lie many more nonconducting cells called "glial" cells. These function as connective tissue to support and protect the neurons and may also play a role in nutritive processes. Although there is as yet no concrete evidence of this, it is also possible that the glial cells support the process of repatterning through the nervous system, helping to make connections between neurons where these links are damaged or broken. There are about nine glial cells to every neuron.

The nervous system consists of the brain and spinal cord, the spinal, cranial, and peripheral nerves, and the special sense organs of taste, smell, hearing, touch, equilibrium, and vision. The brain lies within and is protected by the skull. The cerebrospinal fluid (CSF) flows within membranous linings between the brain tissue and bones of the skull; this fluid acts to lubricate and to absorb shocks to the skull, and also supplies nutrients to the brain. The spinal cord extends down from the lower brain through spaces at the center of each vertebra, which form a hollow channel through the length of the spine. The spinal cord is protected by the vertebrae, the membranous linings, and the CSF that flows around it and within the "central canal" down its length. The spinal nerves come off the spinal cord between each vertebra, one on each side of the spine; these are sheathed bundles of nerve fibers that then separate out like the branches of a tree, radiating to all parts of the body as peripheral nerves. At the level of the second lumbar vertebra, the spinal cord ends; from here on the nerves, no longer protected by the covering of the spinal cord, pass vertically down within the vertebral column until they exit from their respective vertebral openings. These nerves, running parallel within the vertebral column from the second lumbar to

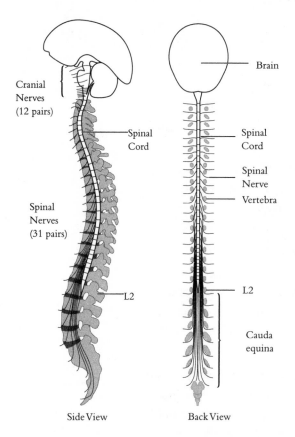

Cranial
Nerves
(12 pairs)

Spinal
Cord

Spinal
Nerves
(31 pairs)

L2

Side View

Brain

Spinal
Cord

Spinal
Nerve

Vertebra

L2

Cauda
equina

Back View

Figure 9.2
Organization of the
central nervous system

the fifth sacral vertebra, form the *cauda equina,* or "horse's tail." Awareness of the fanning of the nerve fibers in the "horse's tail" can give a feeling of added mobility in this frequently too-rigid area. (Fig. 9.2)

Emanating from the brain are also twelve pairs of cranial nerves that are associated primarily with the special senses of the head and the muscles of the face, throat, and neck. One, the tenth cranial nerve, affects the functioning of many of the thoracic and abdominal organs. The sense organs themselves, with their specialized sensory receptors, also lie within the protective casing of the skull, of course with their distinctive openings to the outer environment.

The Brain

The brain itself, which is the highly complex and sophisticated coordinating center of the whole nervous system, regulates all the voluntary and involuntary movements and activities of the skeleton, muscles, and viscera. It organizes sensory input into meaningful perceptions and integrates this into appropriate and purposeful motor responses. Much of its work concerns the control of bodily function, movement, and perception; a later evolutionary development is its propensity for analytical, imaginative, conceptual, and creative thought and activity. This is believed to be almost exclusively an ability of humankind, with our more highly developed cerebral cortex.

The brain can be divided into three main areas: the hindbrain, midbrain, and forebrain (Fig. 9.3). This classification also

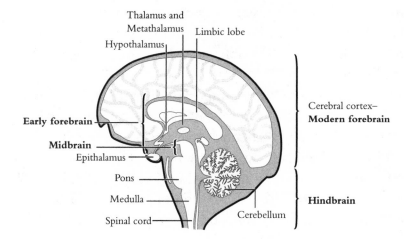

Figure 9.3
The major areas
of the human brain.

reflects the evolutionary development of the brain and of the human species as a whole, the hindbrain being the oldest and most primitive part, and the midbrain and forebrain the most recent and sophisticated. The "mind" of the fish, amphibian, and reptile brain are still accessible to our consciousness. In fact, these "mind" states influence in fundamental ways much of our human experience and behavior.[3] The hindbrain consists of the medulla oblongata (which is continuous with the top of the spinal cord), the pons, and the cerebellum, which consists of two large hemispheres lying at the base of the brain behind the medulla and pons.

The nuclei of the medulla control many of the vital physiological processes of the body that are necessary for survival and are therefore primary in terms of development; these include the functions of respiration, digestion, and circulation. Serious problems with these functions could indicate damage to or dysfunctioning of the medulla. The medulla is also associated with the Mouthing pattern, Pre-Spinal pattern, Spinal Push patterns from the head and tail, and the initiation of head turning. (See Tables 1 and 2 in Chapters Four and Five.)

The pons is also associated with respiration and is a relay center for nerve fibers passing between the two hemispheres of the cerebellum, the cerebral cortex, and the musculature of the body. With the cerebellum it controls the Homolateral Push patterns; it also helps to establish the vertical axis, together with the medulla and midbrain.

The cerebellum is an important center for the control of movement, modifying movement impulses that are initiated elsewhere in the brain, particularly the cerebral cortex. It is associated with equilibrium, postural reflexes, refinement of muscle tone, and coordination in regard to force, time, and space. With the anterior part of the pons it controls the Homolateral Push patterns and balance around, or falling off, the vertical axis.

The midbrain is a small area lying centrally between the hindbrain and forebrain. It is associated with the perception of vision and hearing, protective extension responses, and body-righting reactions. The midbrain controls the Spinal Reach and Pull patterns and the Homologous Push patterns. Anteriorly lie the cerebral peduncles which relay information between higher and lower areas of the brain and the spinal cord. The posterior part of the midbrain consists of the superior and inferior colliculi, associated with visual and auditory reflexes, respectively.

Together the medulla, pons, midbrain, and ancient forebrain form the brain stem, which is essentially a complex extension of the spinal cord. In general, the functions of the brain stem include the control of "respiration, digestion, cardiovascular function, eye movement, equilibrium, antigravity support and various specific stereotyped movements of the body," as well as "the establishing of one's vertical axis or sense of a central core structure."[4]

The forebrain consists of the thalamus, epithalamus, metathalamus, and hypothalamus (the "ancient forebrain") and the two hemispheres of the cerebrum (the "modern forebrain"), together with underlying nerve tracts which connect the various parts. The epithalamus contains the pineal gland, and the metathalamus is a relay center for the visual, auditory, and vestibular areas of the cortex. The thalamus is a relay center for all sensory information (excluding smell) coming from the sense organs to the cerebral cortex (consciousness). Sensations of pain, temperature, and touch are also registered here. The hypothalamus is a control center for the autonomic nervous system and functions to maintain the homeostasis of the internal environment. It links the nervous and endocrine systems, influencing the secretions of the pituitary gland; the mamillary bodies are also situated here. Because of this, the hypothalamus may be experienced as a central place of integration of the mind and body; strong emotions such as rage and aggression are also associated with this center. The Homologous Reach and Pull patterns are associated with the thalamus and hypothalamus.

The thalamus, hypothalamus, epithalamus, and limbic lobe form the limbic system, which is the emotional center of the brain "concerned with the sense of smell and bodily/visceral sensations. It is responsible for subconscious sensory and motor drives, pleasure and nonpleasure, reward and punishment, and approach and withdrawal. It coordinates sensory information with bodily needs and the mind states associated with them."[5]

The cerebrum, or modern forebrain, is the largest area of the brain and is the most highly developed part of the human brain. Here conscious and meaningful perceptions are formed through the evaluation and integration of sensations. It is the center of learning, memory, creative and analytical thought, intellect, imagination, language, and the conscious learning and control of voluntary actions and complex skills. Its freedom to carry out these functions depends on the ability of the lower and more primitive brain areas to integrate sensory-motor processes and regulate the autonomic functions and movement patterns outlined above. The cerebral cortex includes large areas involved in the processing of sensory information and voluntary motor response throughout the musculoskeletal system of the body as a whole; and special centers concerned with speech and the interpretation of auditory, visual, and olfactory sensations. The two hemispheres of the cerebral cortex are associated with the Contralateral Reach and Pull patterns, which bring us to the most highly evolved and complex forms of movement.

Working with the Brain as Organic Tissue

We can work with the brain as organic tissue, using some of the principles outlined in the previous chapter. The different areas of the brain, like the parts of any other organ, can move with or counter to each other to find a more articulate and balanced relationship within the brain itself. By placing our attention on the tissues of the brain, awakening cellular awareness there and becoming sensitive to the sensations of movement or blockage, we can learn to initiate and free the flow of movement. This can have a profound effect upon the freedom and quality of movement throughout the whole body, as well as influencing the physiological and perceptual processes and movement patterns associated with a specific area. Repatterning through the brain and nervous system as a whole can be approached directly through contact with the nerve tissues; or alternatively through the guiding and practicing of movement, perceptual, and expressive activities associated with the area to be stimulated, as described in the chapters on infant movement development and developmental movement therapy.

To contact and repattern through the brain and nervous system directly may at first seem more difficult than working with the other body systems, as our conscious thought processes are so closely related to it. We are trying to sense and organize the system that does the actual sensing and organizing. However, when we contact these tissues at a cellular level, we can perceive and repattern them in the same way, applying the same principles and techniques as for the organ and musculoskeletal systems.

Over recent years the theory that the two halves of the brain are associated with different but complementary processes has become popular.[6] The right brain has been associated with the left side of the body and an intuitive, receptive, "feminine," cyclical, feeling, and artistic "mind." The left brain and right side of the body are more concerned with rational, analytical, "masculine," linear, time- and goal-oriented processes. Although today some people question the validity of this theory, we do nevertheless function within a continuum defined by these two modes of expression, and it can be a useful model to work with. The work of Body-Mind Centering is very much involved with the integration of these two aspects of experience and expression. Exploring our experience of the relationship between the two sides of our own brain, through awareness, sensation, and movement, can be a direct way into this process of integration.

Nervous System Terminology

Structurally, the nervous system includes the central nervous system (CNS), consisting of the brain and spinal cord, and the peripheral nervous system (PNS), nerve fibers that connect all parts of the body with the central nervous system. Peripheral nerves are composed of bundles of either sensory (incoming) or motor (outgoing) nerve cells, or both.

There are motor and sensory neurons, or nerve cells, within both the PNS and CNS. In the PNS bundles of nerve fibers are called "nerves" and collections of cell bodies are called "ganglia." In the CNS they are referred to as "tracts" and "nuclei," respectively. The brain and spinal cord also contain "association" neurons or "internuncial" neurons; these "go-betweens" conduct impulses between sensory and motor nerves within the CNS.

The PNS is further subdivided by both structure and function into two branches, the somatic nervous system (SNS) and the autonomic nervous system (ANS). All of these nerves are out-side the central nervous system.

The SNS directs the voluntary movements of the muscu-loskeletal system; it also conducts sensory messages from the pro-prioceptors of the joints and muscles and exteroceptors in the skin and special sense organs to the CNS. Although called "vol-untary," the process of coordinating these actions can be either consciously or unconsciously directed, and most of the actual sensory-motor processing occurs below the threshold of con-sciousness.

The ANS branches again to include the sympathetic ner-vous system (thoracolumbar nerves) and the parasympathetic ner-vous system (craniosacral nerves). The ANS controls all of those processes that are not normally under conscious and voluntary control. It regulates the activities of the visceral and sense organs, glands, and blood vessels in response to the requirements of the internal and external environment. Some muscles, such as the diaphragm and those of the throat, are innervated by both somatic and autonomic nerves, permitting both conscious and uncon-scious activity in those areas.

The Somatic Nervous System

Sensory nerves within the SNS receive impressions from the external and internal environment and convey this information to the brain (or spinal cord in the case of simple reflex actions). Here, a complex process of "checking" these new sensations against other incoming information and against the memory of previ-ous similar and related experiences allows a meaningful percep-tion of the impressions to be made and an appropriate response to be initiated. Again, some of this processing may be conscious, but most of the sensory integration happens unconsciously. The impulse for the response is coordinated by and transmitted from the relevant areas of the brain, out along the motor nerves to the body parts that are to be activated. With this action further sensations are experienced and new impressions received, which in turn produce new perceptions and motor responses or regulate and refine the activity already occurring. There is thus a continually interweaving cycle of action, perception, and response.

The perception of sensory information itself involves a motor component. We actively perceive by shifting our attention toward certain stimuli and choosing which stimulation we will take in and register, consciously or unconsciously, in the process of organizing sensory information into recognizable and meaningful perceptions. In describing the process of actively perceiving, Bonnie Bainbridge Cohen states that the "'active decision' is usually unconscious, based on previous experience."[7]

The motor activity also involves a perceptual process; we continually receive feedback from our actions and can thus regulate and direct them. This occurs through the activity of "muscle spindle" receptors, "golgi organs" located in the tendons, other sensory receptors, and special areas of the brain. As explained earlier in the discussion on movement development, the vestibular nerves of the SNS, the nerves that perceive movement, are the first to myelinate. Also, the motor nerves myelinate before the sensory nerves; this implies "that one needs to move before one can have feedback about that movement."[8] This feedback stimulates further movement and its refinement or elaboration, and it gives us vital information about who and where we are. "Consciousness," says Lennart Nilsson, "involves perceiving oneself as a separate being, and this requires a continuous supply of sensory input."[9]

Sensory information is received in the SNS through receptors in the joints, muscles, tendons, and ligaments, and through the special sense organs, including the eyes, ears, nose, mouth, and skin. From these stimuli perceptions are formed; perceiving is the way we make meaning out of the sensory impressions we receive. Perceiving is about relating personally to the processes of the inner and outer environment, and it involves choice. In action we align our perceptions with our attention, desire, and intention.

The Autonomic Nervous System

The autonomic branch of the nervous system (ANS), controls the bodily processes that are not normally under conscious control and serves to maintain the physiological balance of the organism in response to continually changing internal and external environmental factors. Its two aspects, the sympathetic and the parasympathetic divisions, are not antagonists as they are often described, but mutually complement, support, and balance each other's functions. Like the SNS, both aspects contain sensory and motor neurons, and therefore both receive sensory information from the body and its environment that then stimulates responses. (Chemical and hormonal "messengers" also play an integral part in the activity of the ANS.) Many anatomists consider the ANS to be only a motor system; in response to this opinion Bonnie Bainbridge Cohen writes, "Whereas, as with the somatic nervous system, there are both sensory and motor neurons involved, the majority of attention [to the ANS] traditionally has been almost exclusively upon the motor aspects. Very little has been written about its sensory pathways and functions. This is due perhaps to the majority of the sensory pathways being invisible to the naked eye. Our study and research at the School for Body-Mind Centering relies greatly upon an extensive autonomic sensory feedback system."[10] In the ANS we receive sensory information through interoceptors in the internal organs, glands, and vessels of the fluid system, as well as through external receptors.

The sympathetic division activates the "fight, flight, fright, or freeze" reactions, which arise in response to threatening or exciting situations in the external environment. Stimulation of this system increases heart rate, respiration, the flow of blood (and therefore energy) to the heart, lungs, skeletal muscles, brain, external sense organs, and the periphery of the body, while decreasing the flow of blood to the internal organs connected to digestion and the urogenital system. A state of alertness and readiness for action is aroused in the senses and body musculature, and the focus is externally directed. If the ANS is not in a state of healthy balance, however, or if the degree of stimulus is beyond the individual's capacity for appropriate and effective response, the reaction may be one of immobility instead of action, of paralysis in the face of danger.

The parasympathetic division produces a complementary condition, increasing the blood supply to and activity of the digestive organs, slowing down the heart beat and rate of respiration, decreasing muscular readiness for action, lowering the alertness of the externally-directed senses, and decreasing activity at the body periphery. It concerns digestive processes, repair and recuperation, or rest. The focus is inwardly directed, with less awareness of and attention to the external environment.

The sympathetic nerves branch off from the spinal cord in the thoracic and lumbar regions; the parasympathetic nerves originate in the brain and sacral regions. (Fig. 9.4) By sensing these nerve pathways and initiating movement through them, it is possible to activate the states of attention, focus, and activity to which they relate. By appreciating their functions we can also determine what kinds of activity would be most helpful to an individual in finding support and balance within the ANS as a whole.

The "mind" of the sympathetic nervous system is outwardly focused, oriented towards activity and the achievement of goals, and helps us to meet the environment. It is connected to the will in action. There is usually an increased level of perception of the external environment through the senses (vision, hearing, smell, etc.). Activities that include vigorous muscular activity, alertness of the senses, quick motor responses (especially in the hands and feet), and that are directed toward achieving a goal are expressions of the sympathetic nervous system. Sports and athletics

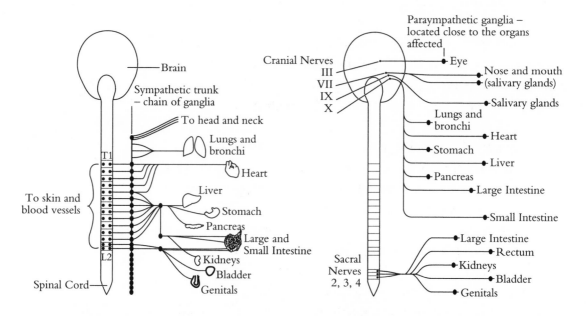

Anterior view – Sympathetic division

Brain

Sympathetic trunk
– chain of ganglia

To head and neck

Lungs and
bronchi

Heart

Liver

Stomach

Pancreas

Large and
Small Intestine

Kidneys

Bladder

Genitals

T1

L2

To skin and
blood vessels

Spinal Cord

Anterior view – Parasympathetic division

Paraympathetic ganglia –
located close to the organs
affected

Cranial Nerves
III
VII
IX
X

Eye

Nose and mouth
(salivary glands)

Salivary glands

Lungs and
bronchi

Heart

Stomach

Liver

Pancreas

Large Intestine

Small Intestine

Large Intestine

Rectum

Kidneys

Bladder

Genitals

Sacral
Nerves
2, 3, 4

The Sympathetic division prepares the body to respond to "fight or flight" situations by:

- Dilating the pupils.
- Increasing the heart and respiratory rates.
- Increasing the flow of blood to the skeletal muscles.
- Increasing blood pressure.
- Causing sweating, and other reactions appropriate to such emergencies.
- Contracting the sphincters of the digestive organs and urinary bladder; relaxing the longitudinal muscles of these organs.

The Parasympathetic division deals with vegetative processes by:

- Stimulating the functions of digestion and absorption through increased secretory and motor activity in the digestive organs and glands.
- Relaxing the sphincters of the digestive tract, and contracting the longitudinal muscles.
- Draining the urinary bladder.
- Drawing blood from the skeletal muscles to the digestive viscera.
- Reducing the heart and respiratory rates.
- Lowering blood pressure.
- Constricting the pupils.

Figure 9.4   The autonomic nervous system, showing organs affected by the sympathetic and parasympathetic divisions. The two divisions complement each other, each stimulating certain functions and inhibiting others. Most organs receive fibers from both divisions.

are excellent examples of activities which give full expression to the sympathetic nervous system; the SNS, which innervates the muscles, is also fully engaged.

The parasympathetic nervous system supports an inner-focused "mind"—one that is self-reflective, digestive, receptive, and process-oriented. It is associated with desire, particularly the desire simply to be, and emotional feelings that are related to organic process. Until about three months old, an infant is primarily occupied with the parasympathetic processes and mind-state. After this time the infant will become more externally directed for longer and longer periods and will respond more readily to the stimulation of toys and other objects in the environment. The parasympathetic nervous system can be contacted and stimulated through activities such as meditation, relaxation, breathing, and sensory awareness techniques.

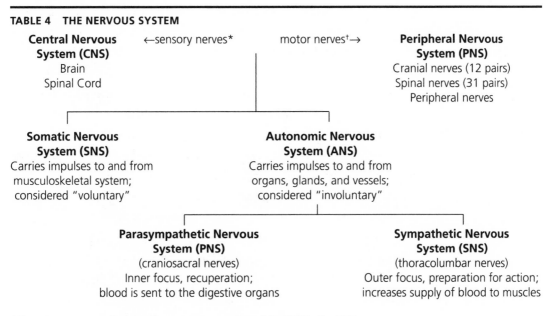

TABLE 4   THE NERVOUS SYSTEM

Central Nervous System (CNS) Brain Spinal Cord	←sensory nerves* motor nerves†→	**Peripheral Nervous System (PNS)** Cranial nerves (12 pairs) Spinal nerves (31 pairs) Peripheral nerves
Somatic Nervous System (SNS) Carries impulses to and from musculoskeletal system; considered "voluntary"	**Autonomic Nervous System (ANS)** Carries impulses to and from organs, glands, and vessels; considered "involuntary"	
Parasympathetic Nervous System (PNS) (craniosacral nerves) Inner focus, recuperation; blood is sent to the digestive organs	**Sympathetic Nervous System (SNS)** (thoracolumbar nerves) Outer focus, preparation for action; increases supply of blood to muscles	

* Sensory nerves carry impulses from sensory receptors of the PNS to the CNS.
† Motor nerves carry impulses from the CNS to organs, muscles, and glands via the PNS.

Balance and Imbalance Within the ANS

Each person expresses, by nature, a unique balance of these two aspects of the ANS. Most of us express most frequently and easily through one or the other, but we of course use both, and each is necessary to the other. The mode that is not expressing needs to be active as an underlying supporting function. For this to happen, it needs to be available to us as an alternative means of expression. Hence, recuperation of our principal expressing mode will usually require some time spent in activities to bring the other mode up into expression. A person who is naturally very outward-focused and goal-oriented needs to spend some time making contact with and developing awareness of her inner world of sensation, feeling, and imagination, through relaxation exercises or meditation, for example. If this is difficult to achieve at first, the process may need to be approached through outward-directed activity, with which the person feels comfortable; the focus can then be gradually guided inward from this starting place. For the person who functions most naturally from a receptive, inner, and process-oriented mind, the senses and the activity of the hands and feet can be stimulated to begin to focus the attention more externally.

As with all other body systems and movement patterns, we do not wish to change a person's innate tendencies but rather give support to their natural ways of expressing by letting them rest and recuperate occasionally; this allows them to function more fully and with less risk of exhaustion. If a person's work or lifestyle requires her to operate in a mode that is against her nature, then she may need to recuperate with the activities that feel most natural to her. For example, a "parasympathetic" type of person working in a very high-powered and product-oriented environment will need to make time for quiet rest and inner-focused activities in order to return to her true nature. Conversely, a predominantly "sympathetic" type of person working in a mode contrary to her nature—for example, a woman taking care of a home and family—will need to engage some of her time in activities focused outside of the home environment that give her a sense of goal and achievement, in order to redress this imbalance. For the majority of people, the dominance of one aspect over the other is not so extreme and they will find that different kinds of recuperative activities are needed at different times.

Bonnie Bainbridge Cohen describes this interaction of the two aspects of the ANS: "Having two harmoniously opposing control systems helps prevent overreaction when the balance point has to shift due to changes in the internal or external environment. Each supports and modifies the other on a wide continuum of attention and function. Alternation of the major active role is also necessary in providing rest and recuperation for the supportive aspect. Together they give us a personal sense of well-being."[11]

One way in which the two aspects of the ANS can become unbalanced is when their direction of focus becomes reversed. If the energy and "mind" with which we need to actively meet the external environment becomes focused inwardly, our internal processes, such as digestion or relaxation, will be approached in an overactive and goal-oriented way. Instead of quietly giving ourselves time to digest our food, we will internally attack it and rush the whole process in order to move on to the next task. Or instead of slowing down to relax, we may be so overly concerned with the end product and the relaxation techniques we are using that we cannot experience the process of relaxation itself—we try too hard to make it happen. Bringing conscious awareness to this pattern, and to bodily sensations such as breathing, can help here, as well as finding a more suitable outlet for this frustrated "sympathetic" energy. Such a pattern may contribute to diseases of the internal organs, such as stomach ulcers, where the lining of the stomach wall is quite literally attacked and "eaten" by excessive acid in the digestive juices.

When a person tries to act upon the external environment

with an inwardly focused mind, there may be a sense of vagueness, ineffectiveness, and lack of real contact with the world and the activity being performed. Attention may again need to be given to both aspects of the ANS. Stimulating the senses, hands, and feet can bring more alertness and attention to and contact with the outer environment; it would also be helpful to support the inwardly focused mind to develop a more creative connection to the inner world. Conscious work with feelings, dreams, and imagery, for example, can yield insight and understanding which help to develop the relationship with the inner self. Both of these two forms of reversal may operate in each of us at times. They are seen in extreme form in both physical and psychological illness when a person becomes stuck in the imbalance and is only able to express from there, unable to adapt and change.

Receptivity and Expression

The nervous system as a whole concerns receptivity and expression; the sensory and motor nerves mediate these two functions. The sensory nerves coming into the spinal cord, from both the somatic nervous system and the autonomic nervous system, enter via the "dorsal root," or back, of the cord. The motor nerves leave the spinal cord at the front, via the "ventral root" on each side. This is the natural flow of input (sensation) and output (activity). (Fig. 9.5) It has been discovered, however, that there are some sensory nerves within the exiting motor nerve fibers that enter the cord via the ventral roots; present knowledge of their function is uncertain.[12]

Bonnie Bainbridge Cohen has applied this information to her observation and experience of what happens within the human nervous system in shock. She suggests that the natural direction of firing within the sensory and motor nerves can be reversed. When a person goes into a state of shock, she is unable to process the information coming in and an appropriate motor

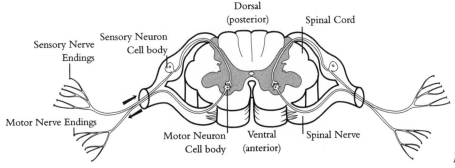

Dorsal
(posterior)   Spinal Cord

Sensory Neuron
Cell body

Sensory Nerve
Endings

Motor Nerve Endings

Motor Neuron   Ventral   Spinal Nerve
Cell body   (anterior)

Figure 9.5
Cross section of the spinal cord showing sensory and motor pathways.

response cannot be made. The situation is more than she is able to deal with or change, and so the information is rejected by the normal sensory process. The nervous system may then reverse its course; information is taken in through the front of the spinal cord, via what are normally the motor fibers. This has the effect of slowing down time; the information may be held in the front of the spinal cord where it is processed more slowly and the motor response is then initiated via the sensory nerves at the back. This reversal acts as a survival mechanism in the face of stimulus that threatens to overwhelm the individual. The patterning will be used until such time as the person is able to consciously accept and process the feelings that she was unable to experience originally. Sometimes they may be held in the spinal cord for a whole lifetime if need be.

The initial response to traumatic events is one of numbness and disbelief. The information is rejected and time stops; there is a freezing that, when extreme, can ultimately lead to death or be witnessed as a kind of living death. A part of us may stay there for a whole lifetime, unconscious to the shock effect, as the trauma hasn't yet been fully experienced. In extreme denial of the trauma, information may be held in the sensory cell bodies lying just outside the spinal cord. When the reversal occurs, information cycles out through the sensory nerves and enters the spinal cord through the motor reversal. Here the shock is experienced. The information can now travel out through the sensory nerves at the back.

Some kind of response can be made, but there is a sense of delay and disconnectedness between the perception and the response—a sense of unreality is experienced. Such altered states of consciousness are commonly experienced in relation to perinatal trauma and have been extensively researched by pioneers such as Stanislav Grof.[13]

It is important to reach the actual point of experiencing the full impact of the shock. This means getting in touch with the emotions or actions that were felt to be unacceptable or overwhelming. If we can hold ourselves within the experience without responding in our habitual ways, it is possible to begin to "de-reverse" this patterning. To do this, the information is held within the CNS for long enough to be processed; this might occur at higher and lower levels of the brain and spinal cord so that consciousness can awaken and energetic changes occur. This requires the safe space of a therapeutic situation or the presence of a close friend or loved one, so that the emotions of the original trauma can be reexperienced and safely contained. Therapy can provide the vessel in which situations and feelings related to the original trauma can arise, but offers a different response from outside so that the client's own patterning may gradually change. Once she can begin to process the experiences that have been held unconsciously within the spinal cord, enough energy will be released to make some motor response possible through the front of the cord, and the de-reversal can begin to happen. Emotional release and body movements or gentle changes in breathing and posture are some of the expressions of this process; they may seem far greater than those the original situation engendered, or they may come quietly as a change in the breathing pattern and shift in awareness.

The healing of a shock response within the nervous system can also be approached through the practice of meditation or disciplines such as yoga and *t'ai chi*, through which we just stay with the experiences that arise within, neither repressing them nor allowing them to overtake us or pull us into habitual reactions. Through accepting our experiences in this way, they begin to dissolve back into a natural state of awareness. The healing effects of this approach are similar, in the long term, to the process of "de-reversal" described above.

The reversal may have occurred at only certain areas of the spinal cord. At such places we are vulnerable to the inrush of stimulus from outside and will feel our emotions strongly here; we experience the information coming into the spine as if through the organs, and its energy can get held there. Actively imagining the natural flow of sensory information coming in through the back of the spinal cord and activity initiated through the front can help the process of de-reversal. Consciously creating and reinforcing this pattern can help to establish it as a prevention against falling into further moments of reversed patterning. This is especially important for the bodyworker or healer, for it is through our own areas of "reversed patterning" that we may be most vulnerable to taking the charge of emotional energy from the client into ourselves. These areas are like wounds in our energy body through which we may be drawn to, attract, and absorb emotional energy in our client that stimulates our own unresolved experiences.

There are other ways in which the natural balance of receptivity and effective expression can be disrupted. These patterns are usually set up in early childhood when we are most vulnerable to the effects of the environment, and we will continue to base our reactions on them throughout life until they are made conscious. When this happens, we have a choice to change the patterns.

If a child does not receive enough of the stimulation appropriate to its developmental age, the nervous system's potential for creating meaningful perceptions and appropriate motor responses will not develop fully. This has been discussed in the chapters on movement development.

A child may also be given sufficient stimulation but not be allowed to reach and have the objects of its attention and desire. There may be perceptions, expectations, and desires, but if these are continually frustrated, motor expression becomes inhibited—the child learns to "look, but don't touch!" and another kind of gap is created between herself and the world. The child may become very sensitive and aware but be unable to express this intelligence effectively. If this situation is extreme and prolonged, the sensory process may also become inhibited as a result. The child needs time to sense, perceive, act upon her environment, and experience the satisfaction of meeting her needs and desires. Only when experiences are complete in this way can learning be fully mastered and natural and effective patterns within the nervous system be established.[14]

Repatterning Through the Nervous System

Within the nervous system we have patterns already determined by previous experiences, through which we will tend to act; they also condition our future responses. As long as we remain unconscious of these patterns, our self-awareness and decision-making ability is limited; they simply cause us to act and respond in predetermined ways. There is no real choice here. At the level of cellular organization, however, there are no such preset patterns and expectations; there is an intelligence and an openness to infinite possibilities. When we contact the nervous system itself at its cellular level, supported by all the cells of the body, change becomes possible. It is at this level that the choice between health or sickness is made.

Often we will work on the nervous system's patterning indirectly, through attention to the other tissues and structures of the body and the movement patterns made through them. When we focus on the tissues of the nervous system itself we can train awareness within it of its own presence, its patterns of movement and perception, and its function of supporting and initiating movement in the body. This includes working with the brain, spinal cord, special sense organs, and spinal, cranial, and peripheral nerves as cellular and organic tissue. As awareness is awakened there, we feel their presence, aliveness and mobility, or stuckness. And as with any other system, this direct perception of the tissue enables us to actively affect their condition and functioning. Movement and energy can be channeled and redirected through the clearest pathways of the brain, spinal cord, and nerves. With our hands we feel for the preferred pathways of movement, for areas of openness and free mobility and for areas of deadness or stuckness. The mind moves, and this can be felt in the subtle passage of motion between the cells of the brain as we lightly touch the head or through the nerve pathways through the body as we extend our mind through them.

Contact with the nervous system requires a very light and sensitive touch. The touch itself reflects the sensing function of the nervous system. It is the inner listening and observing activity that enables us to make conscious our unconscious patterning; through the same process of sensing we can actively direct, through the mind's intent, the repatterning, simultaneously sensing, perceiving, and redirecting the action. Change within the nervous system cannot be forced. It requires making contact with the tissues and perceiving their "mind," their patterning, through the flow of movement within them, then waiting for an opening, a willingness to explore a new pathway. This openness is felt as a space between things; into it the mind can flow and discover new patterns of expression.[15]

There is no effort involved here, but a great deal of sensitivity and careful waiting may be needed to access this space into which we can move. If there is a holding within the tissues, due either to damage, trauma, or lack of use, then we may be looking for a new pathway around the area, awakening a new channel for movement, weaving this way and that around the blockage to stimulate other healthy cells to recognize and take up the movement patterning. Or there may be a need to rest our attention within the stuckness until awareness reawakens there. Then there is a release and the movement can flow through freely. "If there is a wall in front of you and suddenly it is knocked down, you are impressed with a rush of new information. The blockage is a gift. When it is uncovered, it crystallizes an experience you might otherwise have missed. It is very important to experience both the blockage, and the unblockage and the freedom of passage," writes Bonnie Bainbridge Cohen.[16]

When working with someone through the nervous system (either directly, or indirectly through other systems), we follow a principle practiced in the martial arts of *t'ai chi chuan* and *aikido:* to join our own movement and mind-intent with the movement and mind of the other person. Then there is no intrusion, no conflict or force needed. By meeting in this way we merge our mind with their energy's flow and can sense the tiny openings, the spaces, into which our own mind can flow to subtly guide the course of their movement. There can be a great freeing of energy in the smallest of changes, for the mind itself is changing at a fundamental level.

The awareness that enables this sensitive, precise, and finely tuned meeting in the space within which a person is willing to change, to move and be moved, is like the meeting of spirit with spirit.

The "Mind" of the Nervous System

The "mind" of the Nervous System is ever-present in some aspect, reflecting our continually changing modes of interaction with the inner and outer environment. Both inner and outer focus can be either receptive or expressive, and ideally we should be able to move easily among these combinations according to the demands of the environment and our own personal needs.

The essential quality of the nervous system is sensitivity. When our attention is focused on this system and we are acting primarily through it, we experience and express this sensitivity in our movement and awareness as a quality of light and careful attention, looking and listening inward or outward. Through the process of sensing, of inner looking, we gain *in*-sight: we learn to see and know ourselves in a new way, from within. Like all valuable qualities, however, sensing needs to be balanced; if it is used to the exclusion of other processes it can lead to a dryness, a lack of feeling, and disconnectedness from the vitality and reality of life around us. The nervous system is not emotional in nature, but simply communicating information; this means that the process of sensing can be helpful in containing excessive emotions. However, if carried to an extreme, this can lead us to repress and deny these feelings.

Balancing the nervous system are the fluid systems, with their feeling, flowing, and playful natures. A loss of fluidity in the body-mind, where the process of sensing becomes the main mode of expression, places great stress on the nervous system and can create chaos within the unexpressed feelings associated with the fluid systems. If the fluid and nervous systems become seriously unbalanced and out of relationship with each other, a breakdown in one or the other can occur. We will explore this further as we look at the fluids themselves in the following section.

Exploration

To "embody" the nervous system, to ground it in physicality through bringing awareness into its tissues, can energize, release, and strengthen the brain and nerves. Working with the physical presence and substance of the nerves at a cellular and organic level enables us to become conscious of unconscious patterning within the system. Once a pattern has been made conscious, choice is opened to us; instead of acting out of habitual responses we can explore new and more creative ways of initiating movement.

Nervous tissue, like any other tissue in the body, can suffer from too much tension or flaccidity; the quality and tone of the tissue can be positively affected by paying attention to how it functions in movement. The nerves themselves can then provide a tensile support for posture and movement when awareness is brought to them. By releasing tension in the nerves, muscles often relax as a result; what we think of as muscular tension is often tension within the nerves themselves.

1. Begin by lying on your back; slowly and gently rock your head from side to side. Get a sense of your brain within your skull as an organ with weight and mobility; feel this weight shift side to side as you slowly rock your head, imagining the skull as a container filled with water, sand, or beans in a beanbag. Then connect to the weight and movement of the organs in the rest of your body, and let the whole body roll side to side as you rock your head.

Sitting or standing, roll your head around loosely, with your attention still on the brain as an organ; feel how this can give support to the weight and movement of the head. Now connect to the sense of the spinal cord extending down from the brain, through the center of the spine, to its tail. Let the rotations of the head increase so that the spine is curving with it, and feel the brain and spinal cord as one connected whole, supporting this movement from within. You can explore other movements, initiating and supporting from this organic sense of the brain and spinal cord.

2. With the help of a partner, locate in your body the spinal vertebrae and the spaces between each vertebra, one on each side, where the spinal nerves emerge from the cord. Imagine them branching out like long sensitive threads into the body. Trace and

move "through" each spinal nerve, following its direction down and outward, like the sloping branches of a pine tree or the guy-ropes of a tent. Then in stillness feel the whole system of brain, spinal cord, and spinal nerves giving tensile support to the upright posture of the spine and head.

3. From this sense of the spinal nerves, actively imagine and feel the flow of energy going in through the nerves at the back of the cord and out through the front. Feel that all the sensations, sights, sounds, and smells you are aware of right now are coming into you through the back, being sensitive and receptive to them here. Notice any changes this creates in your perceptions of yourself and the space. Begin to move, and feel that this movement initiates through the front of the spine. You can focus either on specific areas of the spine or on it in its entirety.

With a partner, make contact by placing the palms of your hands to theirs, and play with pushing and being pushed, gently at first; again feel that you receive the sensations of the movement and contact through the back of the spinal cord and initiate your responses through the front. Explore this movement with your focus on the different areas of the spine.

4. The next exploration is best done in a group, although you can also work imaginatively with your environment if you are alone. Improvise movement with your attention focused inwardly and: a) be receptive to your own body sensations and movements, letting them inspire and guide your dance; b) feel that you actively initiate movement from within your own body.

Then focus your attention outwardly to your environment and: a) move in response to the sights, sounds, movements and so on around you; be receptive to them, letting them inform your own movement; b) initiate your movement in such a way that you feel you are acting on, affecting, or changing the environment, being yourself the active source of these interchanges.

Your responses to this exploration of inner and outer focus, receptivity and active expression, will give information about both your personal preferences with regard to receptivity and active initiation, and the relationship between the sympathetic and parasympathetic aspects of your ANS. Practicing with the combinations that feel less comfortable and familiar will help to balance the systems.

The Fluid Systems

We saw earlier how all matter shapes itself into spiraling patterns, following the laws of gravity and levity and the tendency for water to seek the form of the sphere (see Chapter Four). Water carries and creates the patterns of organic structure. Through the cycles of creation and destruction of life forms, this seeking to embody the wholeness of the sphere dances with the necessity of change, of breakdown and renewal, a continual process of *re-creation* of the old forms into new life. It is the fluids of the body that mediate these processes of transformation, of health, sickness, decay, death, and rebirth.

As Theodor Schwenk states in *Sensitive Chaos:*

> A sphere is a totality, a whole, and water will always attempt to form an organic whole by joining what is divided and uniting it in circulation. It is not possible to speak of the beginning or end of a circulatory system; everything is inwardly connected and reciprocally related. Water is essentially the element of circulatory systems. If a living circulation is interrupted, a totality is broken into and the linear chain of cause and effect as an inorganic law is set in motion.[17]

Within the human body this interruption of the natural circulation of body fluids causes disorder and disease. Deane Juhan writes:

> Nutrients, oxygen, hormones, antibodies and other immunizers, and of course water, must be delivered to every single cell con-

tinuously if it is to survive and respond the way it should, and all kinds of toxic wastes must be borne away. There is no tissue in the body that cannot be weakened and ultimately destroyed by chronic interruptions of these various circulations.[18]

We must remember also that we do not live in isolation; the whole circulation of water, air, and plant, animal, and human life is an unbroken and interdependent system of which we are a part. Through the circulation of our own body fluids and the air we breathe, we communicate both within ourselves and with the environment outside of us. If this circulation is interrupted or distorted then imbalance is created in our relationship to ourselves, the earth, air, other people, and all the living forms of nature around us. To maintain the wholeness (health) of this living system we need to attend to both inner and outer and the right relationship between the two. The body fluids are the systems through which communication with, and transformation of, both inner and outer environments takes place.

Within the body we find expressions of all the forms in which water circulates on, in, and around the earth. In us there is the great unbounded ocean of fluid in which all the cells are bathed. There are rivers and streams flowing within the vessels of the veins, arteries, lymphatic, and cerebrospinal fluid channels; there are pools and reservoirs and places where the fluid gushes or trickles like springs, waterfalls, or rain. There may also be places where the flow is blocked and the fluid stagnates.

The system as a whole has several subsystems, clearly defined fluids, each with its own chemical nature, consistency, function, pathways of flow, and rhythm of movement. However, it is essentially one system, one fluid, capable of transforming from one subsystem to another—from blood to interstitial fluid to lymph and back to blood again, for example. This transformation happens through membranes, such as the walls of the capillaries (the tiniest branchings of the blood vessels) and cellular membranes that act as boundaries between the systems. It is at the membranes that choices are made—the choice to change from one fluid to another, from one "mind" to another. If we have difficulty in making decisions to change, then our mind is caught at the membrane; instead of an organ of transformation, the boundary can become a too-rigid and impermeable barrier between states of mind and rhythms or qualities of expression. In order to respond to the changing demands of the environment and the task at hand, we need this flexibility of mind.

If, for example, we have been quietly absorbed in reading a book, and are suddenly interrupted by a group of lively children clamoring for attention, we need to make a quick change of "mind" in order to respond to their energy and rhythm. Inability to do this puts great stress on our own system, as we have not found the level of attention, energy, and rhythm that our environment is demanding of us. Expressing from a system and "mind" that is not attuned to the "mind" and energy of the environment can create exhaustion, frustration, and inner tension.

The fluids concern the balance of rest and activity, self-nurturing, nurturing of others, play, laughter, the setting of boundaries and limits in self-defense, active and receptive communication, rhythm, movement, and meditative stillness. Our ability to flow between these states depends on the willingness of our mind to choose to flow with the process of change within and around us. To fully experience the dynamics of life, we need to immerse ourselves fully in whatever is at hand, to go to the very heart of the energy it demands, let ourselves be absorbed in it, moved or silenced by its rhythms. At this point, at the center of involvement, is the potential for change where we can release ourselves fully into another experience of being and activity. As in the yin-yang cycle, the fullness of one aspect often leads naturally into its opposite.

We may by nature be a very "fluid" sort of person, able to express easily through this system and adapt spontaneously. Conversely, we may find this easy and spontaneous flow of rest and activity less familiar and difficult to access. Also, within the various subsystems of fluids, we will usually find some are more accessible than others. Our aim in working directly with the fluids is to bring each one in turn up into conscious expression so that all of their qualities are available; and most importantly, to work with the easy transitioning from one to another, for it is in the ability to make transitions that many of us become blocked. Inability to change means that we remain over-identified with partial aspects of ourselves and of the dynamic expression of wholeness that the fluids reflect. As we bring up our hidden, "shadow" parts into expression, new dynamics of movement are experienced and our habitual modes are allowed to rest, recuperate, and be revitalized. In this way, the "shadow" systems become available both as means of expression and as fuller support for our usual "expressive" systems.

The fluids are emotional in nature; when we express through them there is the vitality and directness of contact that feeling gives to activity. When we move through the fluid systems we feel our movement, as opposed to sensing movement through the nervous system. If we move through a crowded space with fluidity, we have speed, agility, and spontaneity of response available to us; moving while sensing has a more introspective, self-conscious, and less spontaneous, though careful, quality. Here we may be very aware of the movement sensations within our own body, but we are more likely to bump into our neighbor if we try to move too quickly and too close.[19] The fluids, then, are also about fluidity in relationship and expression of all the feeling tones inherent in them. Artistic expression, healing work, and craftsmanship of any sort require a fine balancing of the two processes of "sensing" and "feeling" in activity—balancing skillful control and mastery of technique with heartfelt expression of feeling, rhythm, and dynamics. In this way our personal truth can be expressed through clear and artful form. We also find that in bodywork practice the fluids determine the quality of touch, the "how" of what we do.

We make contact with and initiate movement from the fluids in order to release their natural flow from any inhibitions and blockages that may be present. Wherever this natural flow is interrupted or distorted, there is a holding pattern in the mind, an attachment to a certain way of being; there is also the potential for sickness in the body. As we first begin to re-allow this flow or move into systems that have been closed to us, we often experience strong emotional resistances as we come up against our edges. The moment of change is the point at which we are ready to no longer maintain those barriers but to allow the natural exchange and flow to happen fully and freely. This shift is really quite simply a matter of choice, though often a difficult one that goes against a lifetime of experiencing ourselves in an habitual way. Our choice is to let go. Once the decision is made, the barrier is no more; the membranous boundaries of the fluid systems are then able to function again as organs of transformation.

The release of painful or joyful emotion that may accompany this change is not the block itself, but the freeing of the energy that has been held in. Such changes always have a purifying effect, like clearing a dammed-up river so as to allow the stagnating water to flow again along its natural course. What we observe and experience is that repressed or unexpressed emotional experiences can be "held" within the fluids of the body; it may be that the body fluids attract the charge of emotional energy in a way similar to that of water acting as a conductor for electrical energy. These held emotions can "stagnate" in the same way as a river does when blocked, until awareness of the natural flow is reawakened there; through this awareness the fluids themselves can be repatterned. It is the mind that stops the flow and a change of mind that will release it too. We simply follow the course of nature, the natural pathways and rhythms, with our mind, and the fluids will respond.

This can be done through first studying the anatomy and physiology of the system, understanding the directions and rhythms of its flow, its functions, and the qualities of movement which these suggest. Then through first visualizing and sensing, we can make contact with this system of fluid in our body, its presence, substance, feeling, movement, weight, rhythm. A combination of active imagination, touch, movement, and focused breathing can facilitate this approach.

Alternatively, we can begin with a more active, feeling approach by moving with the qualities, rhythms, and dynamics associated with the specific fluids or dancing to music that will stimulate such qualities. The movement itself will stimulate the flow of the fluids and we can learn from our experience with the different dynamics, or types of music used, which are our "preferred" systems and which systems we find difficult to access. Ideally both the "sensing" and "feeling" approaches are used. Through "sensing" we can gain insight about our patterns and the choices we are making and can make changes more consciously and specifically. In "feeling" ourselves moving, we have the experience of releasing the fluids into their natural mode of spontaneous movement and stillness, rhythm, and dance, which is essential for their integration, and our well-being.

We will now look at the different fluid systems within the body: the venous and arterial flows of blood, the lymph, the cerebrospinal fluid, synovial fluid, cellular fluid, interstitial (or extracellular) fluid, the connective tissue, and fat.

The Blood Circulation

Blood is a fluid tissue that brings life energy to all parts of the body. It flows to every cell, carrying oxygen, nutrients, and the hormones and chemical substances which regulate their activities. The blood serves to nourish and communicate. It also carries away the wastes and toxins no longer needed by the cells, and thus purifies and maintains a healthy internal environment.

If the blood does not flow freely we cannot receive our full potential of energy or fully release the waste substances which may become toxic to the body. The emotional counterpart to this stopping of the flow of blood is that we are unable to nurture ourselves adequately and may store the poisons of pent-up feelings within us. It is primarily through the flow of the blood that emotions held in other body tissues, such as the organs and glands, are moved outwards to expression in movement and communication.

There are two directions of flow within the blood circulation, and we describe this as two distinct aspects of the circulatory system. There is the outward flow from the heart to the body tissues, and the inward flow back to the heart, the central organ in this nurturing, healing, and communicating system. (Fig. 9.6) The heart is a strong muscle that beats tirelessly throughout every moment of life. The heartbeat is an alternation of contraction and release, activity and rest; its ceaseless pulse is the fundamental rhythmic movement of life. It maintains the flow between expansion and contraction, movement and stillness, the cycle of birth and death which is the great activity of the life of the universe.

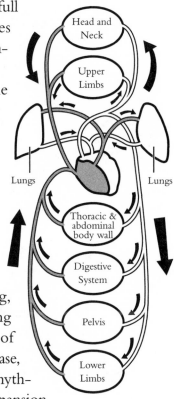

Figure 9.6
Circulation of blood throughout the body.

The Arterial Flow

The movement of blood away from the heart is called the arterial flow. As the heart beats it pumps the blood out through the aorta, arteries, arterioles, and branching networks of tiny capillaries to every part of the body. A subtle wave of contraction through the length of the larger muscular walled vessels supports the initial impulse of the heart, carrying its rhythm through the whole body. The vessels themselves can also inhibit or increase the flow of blood into an area. At the capillaries, the smallest of the blood vessels, oxygen and other nutrients carried in the blood pass through the fine membranous walls to join the surrounding interstitial fluid; from there some of these substances pass through the cells' own membranes into the cells themselves. Carbon dioxide and other waste substances move out through the cell membranes to the interstitial fluid and back into the capillaries, from where the blood continues its flow back to the heart through the network of venules, veins, and vena cava as the venous flow. (We will return to the capillaries shortly.)

From the heart, deoxygenated blood is also pumped to the lungs where carbon dioxide is given off to be exhaled and fresh oxygen is picked up. This flow returns to the heart and the cycle begins again. Freshly oxygenated blood is also pumped into the coronary arteries, which flow directly to the tissues of the heart itself, ensuring its nourishment first so that it can keep up its work of supplying the rest of the body. In circulation through the digestive organs, the blood takes up nutrients to be carried eventually to the cells; passing through the kidneys, the blood is purified of wastes, which are eliminated as urine.

When we move, the arterial flow expresses a lively rhythm of alternating rest and activity, contraction and release of muscles synchronized with the heartbeat. The attention is alert and actively touches the environment; the direction of movement is outward from the heart center, through the limbs and senses toward the external world. The arterial flow expresses vital, outwardly focused, and unselfconscious activity and interaction. The movement is felt rather than pondered on and sensed; it is weighted (bloodfull), and earthy when the substance of the blood is really felt. There is often a playful quality when the flow reaches the extremities and they can interact spontaneously with the environment. The arterial flow is associated with emotional warmth, nurturing, and communication with others.

The Venous Flow

The return flow of blood to the heart center expresses a different quality and rhythm. The rhythmic pulse of the heartbeat no longer exerts such a powerful initiating force for the movement of blood back through the veins; instead the activity of the tissues of surrounding skeletal muscles and organs serves as a "secondary pump" to assist the venous inflow. (More will be said about this pump in connection with the lymphatic system, below.) The veins themselves are less muscular than the arteries and depend on this external pressure to help move the blood along and back to the heart. Valves within the veins open and close rhythmically to prevent a backflow of blood in response to gravity. These actions create a long, slow rhythm, a rising and falling momentum that continually returns back in upon itself. The sustained flow of movement rises and falls within itself, turning and releasing like waves ceaselessly lapping on the shore.

The venous flow expresses the quality of embracing, nurturing, and self-nurturing, a return to and filling of the heart center. There is an inner-directed, feeling quality with the full, earthy, warmth and weightedness the blood gives.

The Capillary Isorings

The places where a continual exchange between arterial blood, cells, and venous blood takes place are called the capillary isorings. Here we experience a moment of rest, a pause before movement and transformation take place. The "mind" in this place of transition is both peaceful and wakeful. (Fig. 9.7)

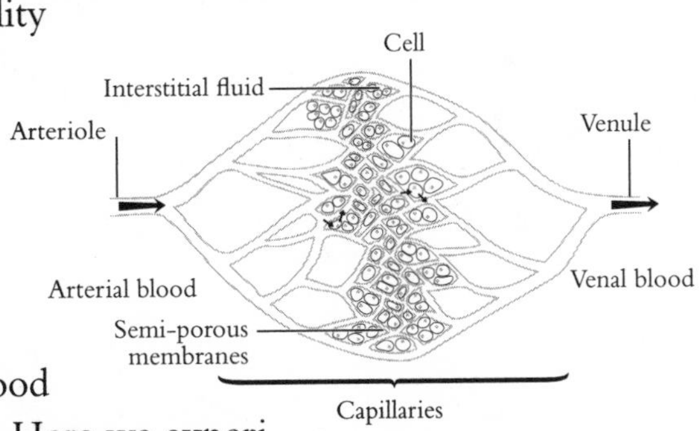

Figure 9.7
Exchange of oxygen, nutrients and wastes takes place between the capillaries and cells, through the interstitial fluid.

275

The Lymphatic System

Closely related to the venous flow in its direction of movement, but very different in quality, is the flow of lymph. The lymphatic circulation forms our fluid system of defense and immunity, together with other tissues and organs including the thymus, spleen, and lymph nodes situated at specific locations in the lymphatic vessels. The vessels form a very delicate silvery network just beneath the surface of the skin and also deep within the body. The lymph flows in only one direction, from the peripheries into the center, and so its flow parallels that of the venous flow of blood. (Fig. 9.8)

Unlike the blood circulation, the lymph is not a continuous system; the vessels are open-ended, webbing into the interstitial spaces throughout the body. Interstitial fluid is drawn into the lymphatic vessels through a special mechanism at the ends of the vessels and is channeled toward the center of the body through them, as lymph. This fluid differs from the interstitial fluid only in its higher concentration of protein, fat, and bacteria. The lymph looks clear and silvery in color and moves at a sustained rate that is much slower than the flow of blood. Because of the nodes' high percentage of white blood cells, the lymph is able to break down and detoxify potentially harmful substances and bacteria, defending us against infection and disease.

The fluid rejoins the venous flow of blood; the small vessels join into larger channels that enter the main veins flowing into the heart in the upper area of the rib cage. The largest channel, the thoracic duct, flows upward along the length of the front of the spine from the cysterna chyli, a lymph reservoir at the level

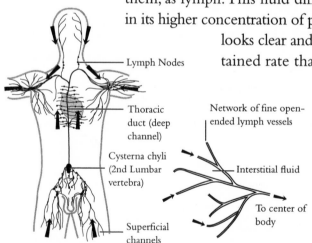

Lymph Nodes

Thoracic duct (deep channel)

Cysterna chyli (2nd Lumbar vertebra)

Superficial channels

Network of fine open-ended lymph vessels

Interstitial fluid

To center of body

Figure 9.8 Flow of lymph throughout the body—deep and superficial channels.

of the second lumbar vertebra. The
activation of this upward flow gives
support to the movement and
posture of the spine and, with the
whole lymphatic network, integrates the limbs with the center. The
integration of the whole system provides a
basis for establishing personal boundaries, which
are associated with physical and psychological
defense.

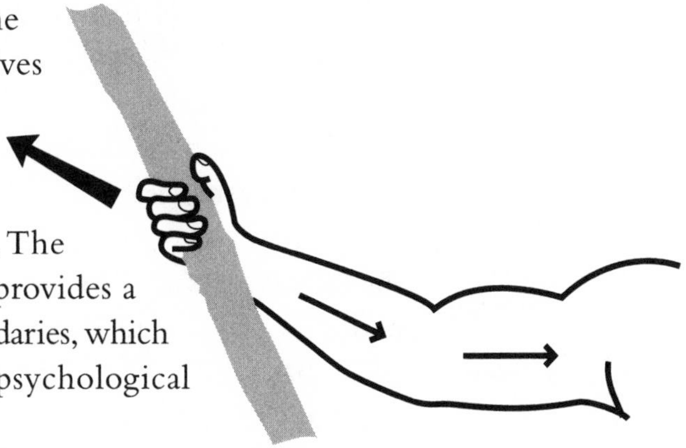

    To feel this setting of boundaries in movement,
imagine the flow of lymph from the very tips of the fingers and
toes, the face, and the eyes in toward your center. The cysterna
chyli at the second lumbar vertebra, the thoracic duct along the
front of the spine, and the heart itself comprise the center in this
case. Imagine activating the lymph's sustained inward flow by
squeezing the muscles of the hands and feet with a firm but fluid
and spongy motion, compressing and releasing rhythmically. This
action is like kneading dough, or the movement the hands make
when revving up a motorbike; the actual movement of your
hands, feet, and the focus of your vision is outward, extending
or pressing toward the limits of your personal space, or kinesphere. This creates space between your center and periphery,
with the lymph acting as a tensile support and connection
between them as it is squeezed along the lymphatic vessels toward
the body center. The boundaries are maintained through the
slow and sustained motion of the lymph in toward the center of
the body, and the counter supporting movement of the vessels
and tissues outward. (Fig. 9.9)

    When actively supporting movement in this way, the lymph
gives clarity of direction and clear presence. This clarity is expressed
through the extremities, which are felt to both support and be
supported by the space. This relates closely to the development
of the Push patterns in the infant, where personal space, bound-

Figure 9.9
The tissue pump—
muscular activity helps
to squeeze the intersti-
tial fluid into the open-
ended lymph vessels, and
supports the flow of
lymph back to the cen-
ter. A steady flow of
lymph helps us to set
limits, or create and sus-
tain boundaries.

aries, and identity first begin to be established. The martial arts also develop and express this system to a high degree, and they are, of course, concerned with self-defense.

The "mind" of the lymphatic system is one of clarity, directness, incisiveness, and the sense of personal power; we literally feel we have a grip on the world, as if holding in our hand a sword, the archetypal symbol of power, to cut through and defend. Games and activities where "weapons" or physical extensions such as sticks, bats, and tools are used in the hands are helpful in stimulating the lymph and the state of "mind" and activity associated with it. If, however, the lymph is overstressed and loses relationship with the other fluid systems, particularly the warmth of the blood, this can lead to rigidity, coldness, intolerance, and humorlessness. We see how an extreme of anything, no matter how valuable or delightful, creates imbalance. The blood has qualities of warmth, sweetness, and empathic communication that without the boundaries and clarity of the lymph may lead to emotional chaos, ineffectiveness, or indulgence and sentimentality. As well as the quality and types of movement we practice, the food we eat directly affects these fluid systems and will create a corresponding quality of expression in our actions and interactions.

Muscular activity, especially initiated from the body extremities, is essential in maintaining the healthy flow of the lymph. The ends of the lymphatic vessels are extremely narrow and the pumping action of the surrounding tissues helps to squeeze the fluid into and along these channels. The subtle cellular activity of the tissues and movement of organs in respiration, digestion, and so on, all contribute to this and muscular activity lends fuller support to the lymphatic system. We call this the "tissue pump" or "secondary pump." Without the activation of the "tissue pump," the lymph flow can become too sluggish, depending excessively for movement on the action of the "primary pump," the heart. This may put added strain on the heart; it needs the support of the activity through the whole body. Simple exercising of the muscles of the hands and feet can almost always be practiced even by a person ill and bedridden; it is especially beneficial when with the movement we imagine and consciously stimulate and direct the flow of lymph toward the center. Massage can also be very beneficial in this respect.

Some techniques of relaxation put emphasis on the outward flow of movement only; they seek to release all tensions in the body using the sensing mode of the nervous system primarily. This might be done by imagining and sensing flows of energy or movement out through the limbs. In excess, this can create an imbalance between the nervous system and the fluids, and can actually lead to a weakening or collapse of the lymphatic system, the inwardly-directed flow of which is not supported. Using the nervous system in this way, we may attempt to create boundaries and defense through it, which can severely exhaust this system. To rebalance, certain necessary qualities of tension and attention need to be recreated through a more dynamic, fluid, and muscular approach to movement; in particular the lymph flow and the rhythm of the arterial blood need to be stimulated into supporting and initiating movement.

Cerebrospinal Fluid

This circulation has already been mentioned in the context of the nervous system. The cerebrospinal fluid (CSF) is a clear and colorless fluid that flows within membranous channels in and around the brain, spinal cord, and spinal nerves; it nourishes, lubricates, and protects the delicate nervous tissues. The brain and spinal cord in fact float within this fluid. It is secreted and stored in four cavities within the brain, the ventricles, from which it seeps slowly into the channels surrounding the brain and spinal cord. (Being aware of these spaces within the brain gives a sense of lightness to the head as it moves and balances.) (Fig. 9.10)

Freedom of movement where the skull and first vertebra

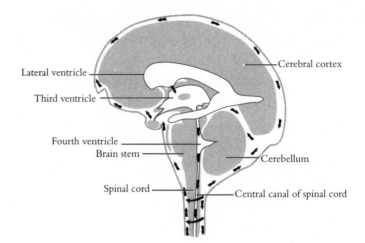

Lateral ventricle

Third ventricle

Cerebral cortex

Fourth ventricle

Brain stem

Cerebellum

Spinal cord

Central canal of spinal cord

Figure 9.10
The flow of cerebro-
spinal fluid within the
central nervous system.

meet is central to the free flow of CSF, as well as blood and lymph vessels and nerves. Some of the CSF also passes through a narrow opening into the central canal of the spinal cord; it is important to keep this space open for the free passage of CSF into the spinal cord. When this flow becomes restricted, as seems to happen particularly in old age, the lower end of this canal may close, and movement of the whole body loses fluidity, youthfulness, and flexibility. A "dryness" sets in, and the head is literally disconnected via this fluid from the rest of the body. The quality of sensory perception also seems to be adversely affected by this blocking of the flow of CSF. In her work with elderly people, Bonnie Bainbridge Cohen has observed that releasing restrictions that inhibit the flow of CSF through the central canal in particular has lessened the effects of aging and senility.

A small amount of the fluid flowing within and around the spinal cord may be reabsorbed into the interstitial fluid; through experiential research, we feel that CSF flows along the length of the nerve fibers to be reabsorbed throughout the body. There is to date no clear scientific evidence of this, but as all body fluids transform from one system to another through the interstitial fluid, it is possible that this does occur. It has also been reported that CSF, or a fluid similar to CSF, is present within the fibers of connective tissue.[20] As osteopath and acupuncturist Fritz Smith describes it:

> The cerebrospinal fluid cushions the spinal cord as well as the brain, and extends for a distance along each of the nerve routes as they leave the spinal cord. Research with the electron microscope indicates connective tissue of the body is hollow. Although it is not generally accepted in Western physiology, many believe that a portion of the cerebrospinal fluid actually leaves the central nervous system via these hollow tubes, spreads throughout all of the connective tissue of the body, and returns to the main fluid circulation via the lymphatic system. According to this viewpoint any vibration within the cerebrospinal fluid would be transmitted throughout the entire body via the connective tissue network.[21]

We can work, through meditation and movement, with this outward flow of CSF, from the ventricles of the brain down through the central canal of the spinal cord and out through the spinal nerves to the body periphery. Imaging and sensing this flow can help maintain a healthy circulation within the whole system and a restful balance between the center and periphery. A considerable proportion of the CSF is continually circulating around the brain and spinal cord; aside from this heaven/earth orientation, the CSF flows only in one direction, from the center outward. This counters the movement of lymph in toward the center.

The movement of the CSF is very slow, almost imperceptible, and it expresses a quality of timeless flow, of suspension in time and space. When the attention settles into the fluid within the ventricles and central canal, we experience a state of meditative stillness and rest. This still quality is carried into the movement of the fluid through the spinal nerves, evoking the light, sensitive, and spacious awareness of stillness in movement that seems to flow into infinity. Through the CSF we contact the "central core of unbounded self."[22]

Simultaneous with the movement of the CSF through the ventricles and channels is a very subtle movement of the bones of the skull and sacrum. This movement first opens the space within the ventricles and channels slightly, allowing for the fluid to fill into them, then closes the space slightly as the fluid flows down and through the central canal. There is a corresponding subtle movement throughout the tissues of the whole body, a barely perceptible opening and closing. It is not certain by what mechanism this simultaneous rhythmic movement of tissues and CSF occurs—whether the moving bones pump the fluid, or the fluid initiates the movement of the bones. Although many theories abound, the source of this deep, subtle, but pervasive rhythm is still one of the mysteries of body-mind process.

The balance of the subtle "craniosacral" rhythm, and flow of CSF, is essential to the health and vitality of the whole person, and its regulation is a primary concern of the cranial osteopath or craniosacral therapist.[23] We can bring our awareness to this circulation through visualizing and sensing the presence of the ventricles and central canal and the CSF flowing through them. This in itself can help to connect us to the meditative state of mind of the CSF; in this restful state the body is allowed to gently return to its own healing rhythms. The treatment of more profound imbalances or restrictions would need professional care, for this is an extremely sensitive and powerful system and changes within it can affect our well-being at the very deepest levels. But regular practice of either sitting or movement meditation (such as *t'ai chi ch'uan*) are an excellent way to maintain the health of this circulation.

To focus the mind a little more specifically into the movement of the CSF, imagine that on inhaling the tissues around the ventricles and central canal are pressed outward and the spaces open and fill with fluid. On the exhalation, imagine the tissues expanding slightly into and narrowing the space, squeezing the fluid down the canal and out slowly through the spinal nerves. (Fig. 9.11) Do this for five to ten minutes. The meditation can then be taken into a slow and sustained movement exploration, feeling the slow seeping of the CSF to initiate movement outward from the center on the exhalation. Such movement expresses the qualities of lightness, balanced suspension in space, sensitivity, and timeless flow that are characteristic of the "mind" or awareness of the CSF.

Together with the lymphatic system, the CSF establishes and maintains spatial tension; these two systems can be explored together to find the balance between their very different but complementary qualities.

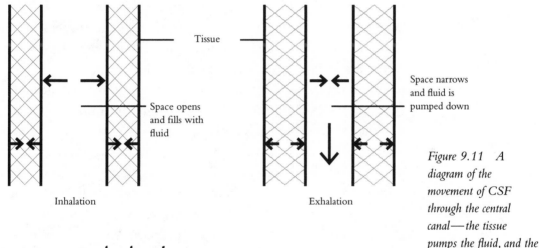

Tissue

Space opens and fills with fluid

Space narrows and fluid is pumped down

Inhalation

Exhalation

Figure 9.11 A diagram of the movement of CSF through the central canal—the tissue pumps the fluid, and the fluid pumps the tissue.

Interstitial Fluid

The interstitial, or extracellular fluid, is the internal ocean in which all the cells and tissues of the body are suspended, float, or swim; it reflects the ocean of the world in which primitive life began and its constituency is in fact similar to that of seawater. During the course of evolution developing life forms have quite literally taken within them a small part of the ocean that was their original external environment. Interstitial fluid has no particular

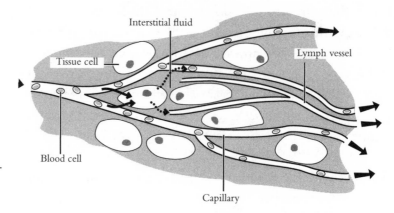

Figure 9.12
The ocean within: interstitial fluid, vessels, and cells.

direction of flow, having no channels or vessels to contain it, but moves constantly in waves, currents, and eddies like the motions of ocean depths, internally connecting the cells and tissues of the body into one ever-changing whole. Diffusing through the membranes of cells, blood vessels, lymphatic, and CSF channels, it continually transforms itself from one system of fluid into another. The interstitial fluid holds, nourishes, and communicates with every part of the body. (Fig. 9.12)

If we follow the arterial flow of blood out to the capillaries and then expand our awareness a little further to the spaces beyond, we meet with the fullness of the interstitial fluid. This gives more weight and power to the movement of the extremities of the body, a sense of presence in space. This sense of fullness gives vitality and power to the movement of the muscles, but with a flowing, undulating, and sensuous quality. The activity of the muscles also stimulates the movement of the interstitial fluid, acting again as a pump which keeps it from stagnating. As described earlier, this activity also helps to return the interstitial fluid through the lymph and veins back to the heart for purification and revitalization. Where energy stops flowing, there the interstitial fluid will also cease to flow; it is in these stagnant pools that the poisons of toxic wastes and unexpressed emotional energy may become trapped. Wherever this happens, the healthy functioning of the cells will be compromised.

We can focus on the interstitial fluid initiating a powerful flow of movement in the muscles; then by shifting awareness into the center of the muscular activity, feel the muscles stimulating the motion of the surrounding fluid. As described in the chapters on movement repatterning, this fluid motion is not restricted to localized areas of the body but ripples and gushes in unpredictable waves of activity throughout the whole body. When we can quiet our thoughts and let go of preconceived ideas about how we want to move, we may begin to feel these inner stirrings as the body's own spontaneous desire to move. By going with these subtle impulses we feel our body move us in unusual and unexpected patterns of flow that are very beautiful to see and are sensuous, satisfying, and healing to experience.

The "mind" of this system is spacious, but almost tangibly weighted, full, vital, and powerful. It contains potentiality, coming and going without restriction.

Cellular Fluid

The cellular fluid is the sphere of fluid contained within the outer membrane of each cell body. Bonnie Bainbridge Cohen describes it as "the center of presence. It is—no coming, no going."[24] Within its holding environment all the essential life processes of cellular activity are carried out. It is the nourishing, nurturing home, the watery ground of life, an oceanic world in microcosm.

There is no specific direction of flow but a continual exchange of fluid through the cell membranes with the interstitial fluid without. This is a process of osmosis stimulated by the chemical contents of the two fluids inside and outside of the cell. The presence of cellular fluid, however, remains constant, only increasing and decreasing slightly with the rhythmic activity of cellular respiration.

Contact with the cellular fluid, like the cell itself, takes us to an experience of being in its full power, simplicity, and omnipotence. The "mind" of the cellular fluid maintains the sense of vitality, power, and spaciousness that we feel in the interstitial fluid, but is more contained. In the cells is a sense of presence that directly touches the depths of our being and the essence of the space around us simultaneously. It is a place of knowing, of body wisdom, from which there is no need to come or go. To be able to act in the world from this base is to be truly present and powerfully at home within ourselves, within our own ground of being.

To feel the cellular fluid, again follow the arterial flow outwards to the periphery of your body; extend your awareness into the surrounding fluid then extend yet further, imagining the fluid diffusing into the cells themselves, each cell expanding to its fullest extent. Pause for a moment, holding yourself in this state of fullness; then follow the cycle back through the interstitial fluid and veins or lymph to the heart. Each stage can be coordinated with an inhalation or exhalation. Repeat this cycle several times until you can feel the energy reaching right to the tips of your fingers and toes. (Remember that where your mind flows, the energy will follow.) Your hands and feet may begin to feel warm, tingling, or "full" as the fluid and the energy begin to move more freely and really fill into all of the cells.

Synovial Fluid

Within the capsules of the synovial joints of the skeletal system (see Chapter Seven) is a viscous fluid, the synovial fluid, which lubricates and protects the joint surfaces and absorbs shocks sustained through movement and the falling of weight through the bones. Like the cellular fluid, it diffuses into and out of the interstitial fluid through the synovial membranes and can transform into blood or lymph to be circulated, purified, and replenished. The synovial fluid is present in most of the joints throughout the body; we might think of them as many small "ponds," feeding into and fed by their tributaries. We can experience and encourage the fluid movement in a single joint, or our attention can follow from one to the another, with each one stimulating the next.

The movement of the synovial fluid is without specific form, direction, or rhythm; where bodily movement is slow and sluggish or bound by tension, the motion of the viscous fluid within the joints may reflect these qualities. However, if we encourage the synovial fluid to flow with a loose, arrhythmic, jiggling quality, we feel that this movement releases the rigidity of the bones and tension within the muscles. When the flow of synovial fluid has become stagnant, we find that a focused "jiggling" of the fluid can be felt to release tension and rigidity in and around the joints. This activity may also stimulate some diffusion of fluids to and from the surrounding interstitial fluid and muscle tissues, which would provide a process of continual renewal of the synovial fluid and a "lubrication" of the muscles, softening their tension into a more fluid flow of action. We experience the "mind" of rigid structure and effort giving way to that of light, carefree flow and laughter. The movement of laughter itself can be felt to stimulate the movement of both the synovial fluid and the interstitial fluid and is very therapeutic in releasing tension and stagnant energy. (Some hospitals are now even providing "laughter therapy"!)

Connective Tissue

Connective tissue, though not generally considered to be a fluid, functions in movement as a circulatory and integrating system, and more than any other system alone it supports the entire structure of the body. It consists, in varying degrees, of semiviscous fluid and cell bodies and fibers; it also varies its balance of tensile strength and elasticity. Some anatomists classify the whole continuum from blood to muscle to fascia to bone as "connective tissue"; all of these tissues are derived from the same layer of embryonic cells. Here we use the term as it is commonly used, to include ligaments and tendons, the periosteum covering bones, and the membranous sheaths, or fascia, surrounding and connecting every cell, muscle, organ, and vessel of the body.[25] The ground substance of connective tissue can be more or less fluid or gelatinous, depending on a variety of factors, such as specific location and function, temperature, movement, and the state of tension, relaxation, and general tone of the body. This fluid is not the same as interstitial fluid, although both surround every cell and tissue of the body; rather it is the medium through which interstitial fluid passes between capillaries, cells, and lymph vessels, and it has distinct functions and a unique composition. The ground substance, cells, and fibers of connective tissue form a continuous sheath surrounding and connecting every cell and tissue of the body, weaving among and linking them into one integrated whole.

The function of connective tissue in bodywork and movement repatterning is a vital one. There can be no change within the body tissues at any level that does not affect the connective tissue sheaths and also no change in the connective tissue that does not effect some change, for better or worse, in the tissues and cells it surrounds. If an organ is collapsed or torqued out of place, a muscle chronically contracted, or a joint misaligned, the supporting connective tissue will also be contracted, pulled, or "glued" together; mobility is then restricted and harmful postural habits become locked into rigid patterns. Cellular functioning is also affected by the quality of the connective tissue.

One of the major benefits of bodywork is in fact an improvement in the quality of the connective tissue. It can better support the other body tissues and cells, as we restore its appropriate consistency and degree of elasticity. Appropriately applied pressure, stretching, and the warmth produced by touch and movement can positively effect the connective tissue, breaking up or dissolving the gluing and solidifying that so often occur when patterns of chronic tension set in. We also find that repatterning movement in a subtle and consciously directed way can help to initiate such changes; as we free movement in specific tissues, the torquing and gluing of the surrounding connective tissue can also be released and the quality of its tone improves. Therefore, as with the nervous system, whichever body system we are working with, we are also effecting change in the connective tissue.

We can also work directly through the connective tissue. When we move in the spaces between the organs, bones, muscles, blood vessels, or cells, we are moving through the connective tissue sheaths. This movement is sensual, internal, flowing, and elastic, like the luxurious stretching of a cat. It flows on and on, rebounds, and flows on again. It gives vital support to the integrity of the whole in both movement and stillness.

Fat

The fat, although not strictly speaking a body fluid, can have certain fluid qualities, and like other forms of connective tissue it can be worked with as such. Layers of fat, or adipose, cells are called adipose tissue; one place we can find this tissue is beneath the dermis, or middle layer of our skin. It is the adipose tissue that insulates the body against cold. Fat deposits cushion and protect many organs; fat exists within membranes, bone marrow, and even behind our eyes. It also forms the myelin sheaths which insulate nerve fibers. The fat is our system for storing energy, to be released when needed in response to hormonal and nervous system stimuli. The synthesis, breakdown, storage, and mobilization of fat are all greatly influenced by hormonal activity; this fact tends to be overlooked in the modern orientation toward calorie-burning cardiovascular exercise. Even very thin people have, and need, body fat; it is essential to health.

Like the fascia, fat can have a mobile and somewhat fluid quality or it can become hard and immobile. Cellulite is an example of this unmoving fat, and like the fascia, it can warm and respond well to touch and movement. It is this hardened and immobile quality of fat that can be problematic and unhealthy. Fat, like every other body tissue, undergoes a process of continual renewal or replacement; this occurs every two to three weeks. However, only fat in its semi-liquid state can be released from the cells and mobilized for use as energy. Fat content builds up when more fat is being stored in the cells than is being released and utilized as energy for the body. Interestingly, the number of cells does not appear to increase in this case; the existing cells simply enlarge. It is well-known that pesticides such as DDT and other environmental toxins may "choose" adipose tissue in which to lodge. While this has led to some positive health consciousness and dietary changes, we must remember that the source of the problem is our increasingly toxic environment, not the fat in which the toxins reside.

Fat is stored as potential power which can either be denied, or accepted and embodied. In modern Western culture fat is perhaps the most repressed, devalued, and feared system of the body. There are always consequences when any group of cells is "exiled" in this way; we might liken this to the silencing of an ethnic minority, in which that voice is lost in the overall balance. We learn to move around our fat, to hold it immobile, to disassociate from it. Those tissues then receive less oxygen, less physical and psychic nourishment, they become less responsive, and the cycle continues. This relationship to the fat is quite recent; Renaissance art and images of dancers earlier in this century remind us of entirely different standards of beauty. We can look to other cultures today to relearn how to embody this system. African cultures, for example, generally support the expression of fat as power which has a sensual and earthy quality, while in many Pacific Island cultures we see a more lilting, buoyant, flowing expression of this system.

When working with adipose tissue we first need to develop

acceptance and a willingness to explore and embody this often unwanted system. Then we need fluidity in order that the potential power that the fat contains can be mobilized and expressed. A fluid touch and a fluid mind help us to contact and mobilize the energy of fat when it has become static, heavy, and hard through repression and negative judgment. Allowing mind to flow through the adipose tissue gives it life and enables movement and change to happen there.

The fat has weight, but this need not feel heavy and solid. The weight of the fat in movement might be experienced and expressed as a billowy fullness of the skin or as a fluid, lively connectedness to the earth. Shaking, jiggling, or vibrating the fat, moving to music such as the deep earthy rhythms of African drums, making vocal sounds, or singing with awareness here are ways to begin to mobilize the fat and experience its buoyant weight and sensuous power. Combining the fat with "true" fluids such as the blood or the CSF will evoke different qualities of it. Fat can also be associated with issues of boundaries and the containment of feelings. In bodywork we can move awareness among the skin, fat, and deeper tissues to explore the roles the fat might be taking in creating boundaries, containing or repressing feeling, and opening to deeper contact.

Fluid Balance

When each fluid system is able to express its unique function and quality of movement and awareness, it becomes both a vehicle of dynamic expression and a support for all the other systems and movement qualities. Activity, rest, and the transitions between these states become easy and fluid; it is in these transitions that much of our energy can be locked or wasted and change becomes an exhausting task. Fluid transitioning enables us to adapt more spontaneously, recuperate more readily, and play more easily within the different qualities of our own being.

If we use a system to provide a function for which it is not intended, we interfere with the natural balance of the whole organism and cannot express that function efficiently. For example, if we use the arterial flow of blood, instead of the lymph, for defense and the setting of boundaries, there will be a hardness and brittleness in the blood flow and an overemotional quality to the boundaries. This hardness may, in time, be reflected in the physical condition of the arteries themselves. Or if we seek the active, rhythmical and outgoing attention of the arterial flow through the venous flow, we will never quite hit the mark and will always feel left behind or late, tired, and missing direct contact with the world outside, for our movement is directed inward with the slow wavelike rhythm and nurturing quality of the flow back to the heart.

The Body's Fluid Outpourings

We replenish our own bodily fluids daily from the greater circulation of the environment in which we live, and we return our wastes to be recycled through the body's natural fluid outpourings. Urine, sweat, menstrual blood, semen, mucus, and tears are ways in which our bodily fluids participate in the greater circulation from inner to outer environment. Our secretions and wastes are necessary to this continual recycling of fluid; in some way each one is a release, a cleansing for us and a gift for the renewal of life. This is often an emotional cleansing or release as well, for the fluids hold our feelings and move them into expression. The flow of fluids into and out of our bodies can also reflect something of our emotional relationship to the world and to life—our feelings and attitudes of connectedness and participation in it and the way in which we receive and give or share energy.

Like laughter, deep or prolonged crying can quite radically affect the fluid content of the body, moving the interstitial fluid which has stagnated in deep areas of the body, as long withheld emotions are released. This has in general a positive cleansing effect; but if excessive, it may also have a destructive effect upon the cells. A deep fluid upheaval, like the breaking of a dam, can weaken the delicate cell membranes, perhaps causing some premature cell deterioration. Such a violent upheaval, often experienced in states of intense grief and trauma, can have a debilitating effect upon the physical body and weaken the sense of psychological boundaries (see Chapter One). Attention to good nutrition and rest, especially the deep relaxation of "cellular breathing," are helpful at such times of profound emotional change and healing; cellular touch or holding helps to restore the integrity of weakened cell membranes.

Exploration

As stated earlier, the fluids can be explored through visualizing and sensing their presence and through feeling them in movement. Through a combination of these approaches, they can be freed to flow in a natural and uninhibited way. Expression of fluidity balances the organizing and controlling activity of the nervous system and allows it to find rest and recuperation. Similarly, the different fluid systems balance and complement each other; when the dynamics and rhythms of each are available to us as means of expression and support we are able to experience more of the wholeness of who we are. Being able to transition easily and readily among the fluid systems enables us to respond spontaneously and appropriately to changing situations, balancing rest and activity, movement in toward our center and expression out in the world.

    1. The more detailed and precise your understanding of the flow of the fluids is, the clearer your connection with them will be. So first take a little time to study the anatomy and function of the different fluid systems, as outlined above. Then try to expe-

*Figure 9.13
Expressing the fluids
freely in improvised
dance movement.*

rience this in your own body through visualizing and sensing their locations, rhythms, and directions of flow, letting your mind travel within the body through these natural pathways.

2. You can also contact the fluids through touch. Begin by making cellular contact with a partner and explore moving your attention to the different fluids. From each fluid you can "resonate" with your partner by simultaneously focusing on that system in her body and your own. If you practice bodywork, you can feel which fluids are most present in your quality of touch. Explore consciously focusing on different fluids to bring in other qualities.

3. It is important that we balance sensing activity by releasing the fluids into a free flow of spontaneous movement. Take the awareness gained from sensing into improvised dance, moving with the quality of each fluid as expressed in its direction of flow, rhythm or pulse, function, consistency, and "mind" state. Now just move and feel the movement, rather than thinking about it. (Fig. 9.13) The following reminders can be used as guidelines for your improvisation.[26]

Arterial Blood: Pulsing; flows out to the periphery; active; rich, full, and weighted; energizing and communicating; rhythm of activity and release; interacts dynamically with the external environment.

Venous Blood: Wavelike; flows in to the center; nurturing and embracing; ebb and flow; rising and falling momentum; rich, full, and weighted; inward movement brings fullness to the heart.

Isorings Fluid: Suspended rest; flowing neither in nor out.

Lymph: Crystal clear fluid; very fine lacey vessels under the skin; a slow, steady flow in toward the center; creates spatial tension between the center and the periphery; defense/immune system; creates and maintains boundaries; crystallization of form; gives clarity and directness to movement; flow is supported by muscle action.

Cerebrospinal Fluid: Clear fluid; acts as a cushion for the brain and spinal cord; slow moving; sense of infinite flow in space and time; suspension between heaven and earth; sensitivity; balance; circular flow within the central nervous system; outward flow through the spinal nerves; meditative rest and stillness in movement.

Interstitial Fluid: The ocean within, home of the cells; movement of the ocean currents; activated by the pumping action of the muscles; spacious, full, and sensuous quality to movement; power and vitality.

Cellular Fluid: Quality of presence; ground of being; "at-oneness"; power, stillness, restfulness; amoeba-like; internal respiration; expansion and contraction of cell membranes; universal life-pulse.

Synovial Fluid: Viscous substance, like egg whites; rebound quality; shock absorbing; arrhythmic; jiggling and throwaway quality; laughter and looseness; unstructured and carefree.

Connective Tissue: Maintains inner integrity of the body; moves through and fills the "spaces between"; luxurious and sensual movement quality, like the stretching of a cat; strength and elasticity; supports every cell and tissue.

Fat: Earthy, sensual, lush; fluid and weighted; light and buoyant; mobilize as soft or strident power; contain as potential power; warmth and humor; big presence; lies under the skin and around tissues and vessels; cushions, protects, and insulates.

When all of these qualities feel familiar, you can practice moving easily and decisively from one fluid to another and playing with combinations of two or more simultaneously. Also, notice which systems you most easily move from and which can give support to, or be gateways into those less comfortable or familiar.

4. Drawing or singing from the qualities of the different fluids can also be creative ways of exploring them.

5. Try dancing to different kinds of music, exploring through your body the rhythms and qualities of the music; try to identify which fluids, or combinations of fluids, are being stimulated and expressed. Express what you are feeling, and have fun!

Conclusion

Chapter Ten

Toward a "Philosophy" of the Body

Today more and more orthodox medical practitioners are acknowledging what the alternative and traditional systems of healing have always known: that there is a self-healing process potentially at work within us, and the mind of the individual has a great influence upon its effectiveness. Since antiquity, physicians and the great thinkers of each era have known that body affects soul and soul affects body. Our culture has just begun to recognize and accept this truth again and even to find many cases of scientific proof and reason for the changes in health that the mind effects. Any vision of health care for the future must include both awareness of the mind's capacity to heal or harm, and acknowledgment of the whole and healthy person each of us potentially is. In this, the individual may begin to take back some of the authority and responsibility for his or her own health which has been laid so completely at the feet of the medical profession. A new vision of health is emerging which is more than protection from nameable diseases; it is a vision of the person as an individual with the freedom and responsibility to make choices and take creative action to nurture his or her own well-being, rooted in each person's own self-healing resources. It is also a vision based on a model of health and potential rather than one of pathology and disease.

The process embodied in the work of Body-Mind Centering serves to empower the individual to effect their own healing and growth toward wholeness. Fundamental to this is the cultivation of self-awareness. As we gain more awareness of ourselves, we can act with greater choice and freedom and begin to take greater responsibility for our own well-being. All genuine healing comes from within; by awakening consciousness to the inner workings of our being, we can release restrictive or destructive habits and allow our innate healing potential to work for us. Any step that helps us take our healing back into our own hands is a step toward personal empowerment and greater respect for our own authority and integrity.

There are also broader cultural implications within the Body-Mind Centering perspective and its model of cellular organization. As we have seen, each anatomical system or structure of the body has its own quality of expression, its own function, needs, and place within the greater whole. Each is equally important to the healthy functioning of the whole organism. The internal relationships of the body systems and functions also offer a beautiful model in microcosm of the relationships between the individuals of a group, the subgroups within society, the nations of the world, and so on. Within the body we find that if one body system or organ, for example, is overused, abused, or denied in some way, the health of the whole person will be compromised. Only when the natural function of each is accepted fully and equally can a healthy relationship among the parts be maintained. Competition or conflict among them results in disorder; each cell and tissue of the body has its own vital part to play and is intended to work in complement with every other part.

And so it is in the larger world: there may be a natural order, but when one group of people is oppressed, exploited, or abused in some way by another more powerful or privileged group, then creative relationships based on mutual support and respect cannot be fostered. Conflict and violence ensue, and social and political stability is gravely endangered. A healthy social or political relationship, following the cellular model, requires genuine respect for the needs and rights of each individual and every group and appreciation of their unique talents and contributions to their society and the planetary population as a whole.

If we can really accept, allow, and nurture our natural order within, there is hope that we may also be able to genuinely accept the strengths, weaknesses, and differences that exist between individuals and groups and allow ourselves to coexist in a relationship of complementarity and harmony without. This change of attitude marks the shift from an awareness dominated by the needs of the ego for personal survival, status, power, and so on, to an awareness guided by the values of the heart and the needs of the greater whole of which the individual is a vital part. Unless we heal the rifts and conflicts within, we cannot hope to effect real and lasting change in the world without, for our world, like our own body, reflects the conditions of the individual and collective mind.

Ancient Wisdom, New Awareness

In one sense, the essence of Body-Mind Centering is not really new, although many of the specifics of it are, I believe, highly original. It might be more true to say that a fundamental process of human development and self-knowledge is being rediscovered, given new form and understanding for the modern Western mind. The many ancient systems of physical yoga testify to the attainment of a profound level of sensitivity and insight into the processes of the human body-mind and energy systems in gifted and enlightened men and women of the past. The word "yoga" literally means "to yoke"; in its deepest sense it means union. Like other systems of physical yoga, and I include here all body-mind practices such as meditation on the breath, hatha yoga, the martial arts, Sufi ecstatic dance, and so on, Body-Mind Centering practice is concerned with developing awareness of the movement of mind and energy; with experiencing the connections that exist between body, mind, and spirit; and with clarifying and deepening these connections and using them creatively, therapeutically, or toward a path of spiritual development.

We could say that such practice is a bridge by which we may consciously connect the processes of mind and body, and link these inner processes with the forms of expression they take in the world. Unlike the ancient systems of yoga that are by now more widely known, this work is still in a very youthful and exploratory stage of its development and its direction of growth is hard to predict. But it does offer something quite unique in giving to some of these processes a new language that has developed out of, and speaks to, the mind of modern Western culture, while also being grounded in and nurtured by the perennial wisdom of both East and West.

In comparing Body-Mind Centering practice to yoga I do not wish to claim that it is in itself a spiritual path; but it has been for me personally, and I think for many others as well, a door opening into the deeper levels of ourselves and others and the universe in which we live, helping us to place our individual lives within a greater context of spiritual life. Because the work is about learning of an experiential nature, it is only possible for me to write about it from a perspective of personal understanding and observation. I hope that I have also remained true to the essential nature of the vision, process, and principles of Body-Mind Centering.

The new consciousness now emerging in our culture includes a balancing and synthesizing of "masculine" and "feminine" principles and values. This means, when we come right down to our personal as well as cultural experience, a reinclusion of the qualities and values of the feminine and a reassessment of the way the masculine mode is expressed. Modern Western philosophical and religious beliefs and attitudes have, for centuries now, denigrated the feminine qualities and the body itself as troublesome and unholy; with this has come a loss of respect and proper care for the earth, for women, for motherhood and so for children too, and for the physical, feeling, and feminine nature in all of us. Relearning to feel at home in our bodies and on the earth is part of this process of revaluing the "feminine," and redressing this imbalance in our culture.

The advancement of industrial and technological cultures have also separated us more and more from the earth and the free movement of our bodies. Most people today perform within only a very limited range of the movements that their bodies are capable of doing. This can cause a gradual stagnation of energy and loss of expression, vitality, and awareness. The emphasis on progress and achievement has caused us as a culture to lose touch with our roots in the earth and with the receptive and intuitive ground of our being. The earth herself is our teacher. If we pay attention to her messages, the earth can guide us in taking care of both herself and ourselves. To quote Alan Bleakley:

> [H]er teachings remind us to look within to our "first circle" and teacher—to the health of our own "earth"—the body itself. She teaches of looking inside to a core, a place of truth and healing; and she teaches mainly about death and rebirth, life-in-death and death-in-life.[1]

The Dharma of working with the body involves healing these rifts within our personal and cultural awareness, to re-own the wisdom and love that are our own true nature. As we are able to heal the splits within ourselves, we will be able to help others heal themselves also. Gradually, the cultural mind will begin to change as the reverberations of each individual act of awareness and caring ripple outward. Learning through our own deep and personal experience is a path toward embodying the truth of who we are and developing the genuine confidence that enables us to express that truth in our lives. As we come home to ourselves we will begin to learn the simple quality of being kind to ourselves, to others, and to the earth. We contact our own heart and the hearts of others.

Evolving a Language for Body-Mind Experience

Body-Mind Centering work offers a means by which we can gain a more intimate knowledge of ourselves. It is a process rather than a technique or set system to be learned and followed. It would be impossible in a work of this size to present in full the vast and still growing body of information, knowledge, and insight that Body-Mind Centering encompasses. I have attempted here to give a framework and some guidelines for the practice of an approach which has almost limitless possibilities. The learning we gain from applying this process is unique to each person; for each of us it reveals our subjective and experiential truth. The descriptions of the various body systems and patterns and their associations need to be viewed within this light—not as rigid formulae, but as insights gained through many people's experiences over a period of time. Further and deeper exploration may in time refine, elaborate, or even radically alter these insights. As the individual and cultural mind evolves, perspectives shift and new levels of experience and knowledge can be reached.

I believe that this work itself reflects such a shift in the cultural mind that is now taking place, and as such is both a work born out of and for this particular moment in human evolution. Body-Mind Centering offers a radically different way of experiencing the body-mind relationship to that which has dominated our culture for centuries. Because of progress in scientific research, we now also have a more sophisticated and accurate model of the workings of the human body and brain than former cultures may have had. A new language for the experiences of body and mind is beginning to evolve, a language that reflects the emergence of a new level of conscious awareness in the psyche of the culture at large. Necessity and the current availability of physical, psychological, and spiritual techniques for personal growth and development are bringing the possibility of such increased awareness into the lives of many, rather than to only a few devoted or gifted people.

It is not easy to formulate this new language, for it must reflect a shift in perception and consciousness from viewing the body only as "something out there," a material object to be studied scientifically, to experiencing it as a living subjective reality, the embodiment of our innermost self through which we experience and learn.

What I have attempted to present here are some of the basic elements of Body-Mind Centering, one such still-evolving language. The words themselves will never quite convey the experience; they are the map, not the ground itself. But they can act as guides into the exploration of the actual territory. And perhaps by feeling your way into the words—the descriptions of the body and its movement—they may convey some hint as to the experience itself. I noticed, for example, that as I wrote about the cell, the chapter had no clear linear development but tended to revolve around, in, and out of a central idea, in the same way that a cell's structure is a sphere organized around a central nucleus. The chapters on nervous and muscular systems, on the other hand, took on a much more linear and logical form of organization, reflecting the particular structures and functions of these subjects. The body structures and systems actually define the quality of their own maps in language, just as they define the qualities of movement and mind.

Although some of the ideas presented here may seem unusual to you, it is also just as likely that you won't see this work as either new or strange but that it will touch upon many experiences familiar to you. It offers a *language* for what we already know through the innate wisdom of the body, and for what we express, consciously or unconsciously, through our actions. By defining our experiences through language in this way, these experiences may be remembered, brought to consciousness, clarified, seen, or articulated in a new way. The act of bringing to consciousness and into language that which was unconscious, unknown, is in itself a process of empowerment.

Essentially we are studying nature and life, rather than pathology and death, as traditional anatomists and medical scientists have often tended to do, basing much of their knowledge on the study of cadavers or the sick and injured. Therefore we may come to slightly different conclusions when we study the body moving and animated by the forces of life and health. We are all the time exploring ways to allow this nature to *be*—allowing the natural patterns and flows of movement, the natural process of development, the nature of aligning ourselves with earth and heaven, and the nature of the process of change. This nature is our potential, and as such it exists within our awareness at some level of consciousness. Movement experience "reminds" us of what we have forgotten, both individually and culturally. In studying the physical body in motion, we are also studying the mind or consciousness that illuminates and is sourced through the physical body.

As we explore the developmental movement patterns, for example, we remember that we once experienced moving and perceiving like a fish or reptile. This information is encoded within our genes and actually calls for expression in some form, in order for our health and potential to develop fully.[2] People of earlier and so-called "primitive" cultures, whose lifestyles were more physical and in harmony with the earth than those of modern society, gave expression to much of this movement potential in their daily activities—hunting, climbing trees, diving to catch fish or collect seafood, cultivating crops by hand, carrying water jars or baskets on the head, engaging in primitive warfare, and in ritual dances that reenacted these activities. This great variety of actions gave a natural outlet for the body's need to continually express and strengthen its basic physical and psychic potentials. It seems that our own culture is now recognizing this need, as is seen in the recent increased interest in fitness, exercise, sports, dance, the martial arts, and so on. Culturally speaking, the body, in its wisdom, is recognizing the imbalances we have created and is reasserting its own needs.

As we go inward to the source and sensations of our movement and from this deep source create our own personal dance, we also find that we rediscover dance forms characteristic of other times and cultures. It is as if we also carry within us our collective cultural history, its rich source of ritual and artistic expression.

Another way that this work may be recognized is as a language for experiences we have had through dance or other movement education disciplines. Each system of movement or style of dance emphasizes a particular combination of body systems and may also utilize the energy of specific glands or organs. This energy is reflected in both the form of the movement and in the quality of its expression. To use two very different examples: some of the earthy, rhythmical, full undulating movements typical of traditional African dances express particularly the arterial blood flow, muscles, organs, and the glands of the lower body centers; and, the "Release" techniques of postmodern dance tend to emphasize the clear, flowing lines of the skeleton, ligaments, nervous system, and CSF flow. Within this there will of course be individual variations, but we can observe general tendencies and recognize that particular styles of dance will stimulate certain body systems primarily.

In bodywork and massage, we may work primarily through certain combinations of systems; this will depend on the technique being used, the practitioner's own qualities, and the client's needs. An understanding of the body systems is helpful in clarifying how we work. It gives an added perspective from which we can choose at which level or body system we might approach the symptoms being presented by the client. By shifting our focus in the body the quality of our touch also changes and can then affect more directly the tissues involved. Thus Body-Mind Centering also offers a body-based language for the art of touch.

Of course, to practice these techniques with clients or students, professional training is required. The reader wishing to gain a deeper experience and understanding is recommended to contact a qualified Body-Mind Centering practitioner or teacher. (Information on where to contact qualified practitioners and professional training programs in Body-Mind Centering appears on page 333.)

A "Philosophy" of the Body

I would like to end on a philosophical note. The word "philosophy" comes from the Greek words *philos,* meaning "love," and *sophia,* which means "wisdom." Sophia is a goddess; she represents the creative feminine spirit, intuition, and the wisdom born of the dark womb of life and death. Her qualities are relationship, feeling, inclusiveness, holding, and nurturing. Philosophy, then, means "love of wisdom," and a philosopher would be one who is both loving and wise. In Tibetan Buddhist tradition, these qualities of loving compassion (*karuna,* said to be "masculine") and wisdom (*prajna,* said to be "feminine") are described as the two wings of a bird that working together lead us to enlightenment. We cannot realize our wholeness without developing our capacity for both love and wisdom.

Poet and philosopher Robert Bly once described a further meaning of the root word *sophia* as being craftsmanship, skill with the hands—the skill of the weaver who combines threads into new patterns, of the seamstress who sews things together, or of the maker of sails by which the ship makes its journey through stormy seas.[3] If we understand the meaning of philosophy to be the love of wisdom and craftsmanship, it is a far cry from the academic, rational, and logical processes that have come to be known as "philosophy" in Western culture. Yet this definition is closer to the work of all those who seek to bring healing to others through the wisdom, caring, and skill developed from their own direct experience and learning.

Bonnie Bainbridge Cohen is one such person, and I would like to express my appreciation for her contribution to this great work. I thank her also for the love and wisdom with which she continually teaches, inspires, and heals. Through insight, words, and the touch of loving hands, she has helped many people on their journeys back home.

Coming back home means returning to ourselves, discovering our wholeness and our inherent good nature. Being and feeling at home in our bodies is essential to this discovery of our basic health and well-being. When we begin to explore, with love, respect, and the skill of sensitive and trained hands, the wisdom revealed by the body-mind, we discover for ourselves a way of returning a little closer to home and to a truer philosophy of life. And when we learn to love the wisdom of life as it expresses itself in nature, we begin to see that our symptoms are pointing the way—are in fact the way itself. The symptoms of our disease and distress are "Home en route,"[4] the signs that reveal both our obstacles and what is needed in order to move beyond them and rediscover our essential wholeness, our innate good health and integrity.

When the body is deeply, lovingly touched,
heart opens, soul awakens.
As the heart is opened, spirit moves.
As the spirit moves
the source of the dance is felt—
we are touched by the dance.
We are called.

Notes

Introduction

1. Hexagram No. 24, *I Ching: The Book of Changes,* Richard Wilhelm, translator. (London and Henley: Routledge & Kegan Paul, 1968), pp. 97–98.

2. Dianne M. Connelly, *All Sickness is Home Sickness* (Columbia, Maryland: Center for Traditional Acupuncture, 1986).

3. This work is described in Mary Fulkerson, Nancy Udow, and Barbara Clark, *Theatre Papers* (Totnes, Devon, England: Department of Theatre, Dartington College of Arts, 1977).

4. Bonnie Bainbridge Cohen, quoted in Nancy Stark Smith's interview "Moving from Within," in *Sensing, Feeling, and Action: The Experiential Anatomy of Body-Mind Centering* (Northampton, Massachusetts: Contact Editions, 1993), p. 11.

5. Cohen, "Moving from Within," p. 8.

6. In Body-Mind Centering work, the term "body system" means an anatomical system of the body, such as the skeletal or muscular system.

7. Candace Pert describes new research on the interconnectedness of mind, body, and emotions and the essential role that neuropeptides play in her article "The Wisdom of the Receptors: Neuropeptides, the Emotions, and the Bodymind," in *Advances: The Journal of Mind-Body Health,* Vol. 3, No. 3, Summer 1986, pp. 8–16.

8. Pert, "The Wisdom of the Receptors," p. 9.

9. Pert, "The Wisdom of the Receptors," p. 14. Emphasis added.

10. We might remember here that science itself is never with-

out an element of subjectivity and relativity, as it is influenced by the knowledge, preoccupations, attitudes, beliefs, feelings, and so on of the scientist and the culture in which he or she lives. Science is also inspired by the scientist's creative imagination and intuition. Science and rationality are not able to speak of ultimate truths but only of the relative truths to which our collective intelligence has thus far gained access.

11. Marion Woodman, *The Ravaged Bridegroom* (Toronto, Canada: Inner City Books, 1990), p. 43.

12. The theory of psychological "subpersonalities," and its application in therapy, has been developed and made popular through Assagioli's psychosynthesis methods, among others. Information can be found in Roberto Assagioli, *Psychosynthesis* (Wellingborough, England: Turnstone Press Ltd., 1975), and Piero Ferucci, *What We May Be* (Wellingborough, England: Turnstone Press Ltd., 1982).

Chapter One

1. Ken Wilber gives a comprehensive study of the different levels and transitions in the development of human consciousness in *The Atman Project* (Wheaton, Illinois: The Theosophical Publishing House, 1980).

2. Lennart Nilsson describes this process clearly and simply in *Behold Man* (Boston: Little, Brown & Co., 1974), pp. 28–30.

3. Rupert Sheldrake describes his theory of morphic resonance, by which the collective memory of the species may shape the individual's physical and social development, in *The Presence of the Past* (London: Collins, 1988).

4. Experience and research with dolphins, for example, is challenging the long-cherished notion that humans are the sole possessors of "higher" consciousness.

5. Master T. T. Liang, *T'ai Chi Ch'uan for Health and Self-Defense* (New York: Vintage Books, 1977), p. 70.

6. The Simontons are pioneers in the field of research into emotional and stress-related factors in the development of cancer and in approaches to healing. See O. Carl Simonton, Stephanie Matthews-Simonton, and James L. Creighton, *Getting Well Again* (New York: Bantam Books, 1980).

7. Diane Connelly, *Traditional Acupuncture: The Law of the Five Elements* (Columbia, Maryland: Center for Traditional Acupuncture, 1987), p. 3.

8. See Wilber, *The Atman Project,* for a discussion of this.

9. Bonnie Bainbridge Cohen, Ruth Leeds, Linda Kalab, Susan Peffley, and Kay Wylie, *The Skeletal System: Manual for a Workshop in Body-Mind Centering* (Amherst, Massachusetts: The School for Body-Mind Centering, 1977), p. 3.

10. Deane Juhan, *Job's Body* (Barrytown, New York: Station Hill Press, 1987), p. 29.

Chapter Two

1. Ken Wilber, *The Atman Project* (Wheaton, Illinois: The Theosophical Publishing House, 1980), p. 83.

2. Bonnie Bainbridge Cohen, "The Action in Perceiving," in *Sensing, Feeling, and Action: The Experiential Anatomy of Body-Mind Centering* (Northampton, Massachusetts: Contact Editions, 1993), p. 115.

3. See Marylou R. Barnes, Carolyn A. Crutchfield, and Carolyn B. Heriza, *The Neurophysiological Basis of Patient Treatment, Vol. 2: Reflexes in Motor Development* (Atlanta, Georgia: Stocksville Publishing Co., 1978); Mary R. Fiorentino, *A Basis for Sensorimotor Development—Normal and Abnormal: The Influence of Primitive Postural Reflexes on the Development and Distribution of Tone* (Springfield, Illinois: Charles C. Thomas

Publishers, 1981); and Bonnie Bainbridge Cohen, "The Alphabet of Movement (Part I & Part II)," in *Sensing, Feeling, and Action: The Experiential Anatomy of Body-Mind Centering* (Northampton, Massachusetts: Contact Editions, 1993), pp. 122–156, and *The Evolutionary Origins of Movement* (Amherst, Massachusetts: School for Body-Mind Centering, 1986).

Chapter Three

1. See Stanislav Grof, *Beyond the Brain* (Albany, New York: State University of New York Press, 1985), and *The Adventure of Self-Discovery* (Albany, New York: State University of New York Press, 1988).

2. Joseph Chilton Pearce, *Magical Child* (New York: Bantam Books, 1980), p. 52.

3. See Frank Caplan, *The First Twelve Months of Life* (New York: Bantam Books, 1978); and Ronald S. Illingworth, *The Development of the Infant and Young Child* (New York: Churchill Livingstone, 1983).

4. See *The Illustrated Encyclopedia of the Animal Kingdom* (Danbury, Connecticut: The Danbury Press, Grolier Enterprises Inc., 1968).

5. In current practice at the School for Body-Mind Centering, the term "Push" pattern has been replaced with "Yield and Push" pattern. For the purposes of this book I have retained the original terminology.

6. "Phylogeny" refers to the species, "ontogeny" to the individual.

7. Although the inchworm is not a vertebrate animal, I have used it as an example here because its movement, levering against a solid surface, best illustrates the movement of the Spinal Push pattern. This usage is a personal variance from current School for Body-Mind Centering teaching, which

uses the example of the fish, a vertebrate creature, to illustrate this movement pattern.

8. The concept of the basic planes of movement was first developed and applied to movement observation and education by Rudolf Laban. See Irmgard Bartenieff and Dori Lewis, *Body Movement: Coping with the Environment* (New York: Gordon & Breach Science Publishers, Inc., 1980).

Chapter Four

1. I include here the element of "space" in which the four elements of earth, fire, water, and air have their existence; this is the system used in Buddhist psychology.

2. Theodor Schwenk, *Sensitive Chaos* (London: Rudolf Steiner Press, 1965), p. 13.

3. *Temple Fay, M.D.: Progenitor of the Domain-Delacato Treatment Procedures,* James M. Wolf, editor (Springfield, Illinois: Charles C. Thomas Publications, 1968), p. 117–131.

4. See Madeleine Davis and David Wallbridge, *Boundary and Space: An Introduction to the Work of D. W. Winnicott* (London and New York: Penguin Books, 1983).

5. The "kinesphere," as personal space, is another concept developed in movement studies by Rudolf Laban.

Chapter Five

1. See Carolyn Shaffer's article on the work of Emilie Conrad-Da'oud, "Dancing in the Dark," in *Yoga Journal,* November/December 1987, pp. 48–55, 94, 98.

2. See Margret Mills and Bonnie Bainbridge Cohen, *Developmental Movement Therapy* (Amherst, Massachusetts: The School for Body-Mind Centering, 1979).

3. The information presented here on the brain and developmental patterns is based on earlier research by Bonnie

Bainbridge Cohen. Further research has been done at the School for Body-Mind Centering since then, as Ms. Cohen continues to refine this system. The information given here should be taken as a general guideline to Body-Mind Centering principles, and not in any way a final version of what is in essence an ongoing process of research and articulation. For more information, please refer to Bonnie Bainbridge Cohen's recent writings.

4. This information is based primarily on the writings of Bonnie Bainbridge Cohen, in particular "The Neuroendocrine System" (Amherst, Massachusetts: The School for Body-Mind Centering, unpublished), and amendments given to me personally by Bonnie Bainbridge Cohen.

5. Madeleine Davis and David Wallbridge, *Boundary and Space: An Introduction to the Work of D. W. Winnicott* (London and New York: Penguin Books, 1983), p. 65.

Chapter Six

1. See "Sensing, Feeling, and Action: An Interview with Bonnie Bainbridge Cohen" by Nancy Stark Smith, in *Sensing, Feeling, and Action: The Experiential Anatomy of Body-Mind Centering* (Northampton, Massachusetts: Contact Editions, 1993), p. 64.

2. Cohen, quoted in "Sensing, Feeling, and Action," in *Sensing, Feeling, and Action,* p. 64.

3. Richard Moss discusses the issue of boundaries and unboundedness in relation to cancer and schizophrenia in *The Black Butterfly* (Berkeley, California: Celestial Arts, 1986).

Chapter Seven

1. Deane Juhan, *Job's Body* (Barrytown, New York: Station Hill Press, 1987), p. 34.

2. See, for example, Wynn Kapit and Lawrence M. Elson, *The Anatomy Coloring Book* (New York: Harper and Row, 1977), pp. 102, 107; Edwin B. Steen and Ashley Montagu, *Anatomy and Physiology, Volumes I and II* (New York: Harper and Row, 1959), Vol. I, p. 35; Vol. II, p. 80; and Lennart Nilsson, *Behold Man* (Boston: Little, Brown & Co., 1974), p. 52.

3. Juhan, *Job's Body,* p. 34.

4. Juhan, *Job's Body,* pp. 107–108.

5. Bonnie Bainbridge Cohen, Ruth Leeds, Linda Kalab, Susan Peffley, and Kay Wylie, *The Skeletal System: Manual for a Workshop in Body-Mind Centering* (Amherst, Massachusetts: The School for Body-Mind Centering, 1977), p. 5.

6. Bonnie Bainbridge Cohen, et al., "Joints and Ligaments." Unpublished manuscript. (Amherst, Massachusetts: The School for Body-Mind Centering, 1982), p. 1.

7. Cohen, et al., "Joints and Ligaments," p. i.

8. See I. A. Kapandji, *The Physiology of the Joints, Volume 3: The Trunk and Vertebral Column. Second Edition* (London and New York: Churchill Livingstone, 1974).

9. Cohen, et al., *The Skeletal System,* p. 5.

10. See Carmine D. Clemente, *Anatomy: A Regional Atlas of the Human Body, Third Edition* (Baltimore, Maryland and Munich: Urban and Schwarzenberg, 1987), plates 134, 614, 616, and 617.

11. Cohen, et al., "Joints and Ligaments," p. i.

12. In Laban terminology, movement of the forearm on the upper arm is named a "distal movement"; Body-Mind Centering discriminates the same movement more finely in describing it as a "proximal *initiation* of movement."

13. Cohen, et al., *The Skeletal System,* p. 4.

14. Juhan, *Job's Body,* p. 98.

15. Cheng Man-ch'ing, *T'ai Chi Ch'uan: A Simplified Method of Calisthenics for Health & Self-Defense* (Berkeley, California: North Atlantic Books, 1981), p. 8.

16. Kapit and Elson, *The Anatomy Coloring Book,* p. 9.

17. Bonnie Bainbridge Cohen, 1990 Addendum to *The Skeletal System*, p. 2.

18. Juhan, *Job's Body*, pp. 113–114.

19. The term "the small dance" was coined by Steve Paxton, originator of the dance form "Contact Improvisation." He describes his work in *Theatre Papers: First Series, Number 4* (Totnes, Devon, England: Dartington College of Arts, 1977).

20. Deane Juhan gives a detailed and in-depth description of muscle function in *Job's Body*, pp. 109–144; 183–244.

21. Clem W. Thompson, *Manual of Structural Kinesiology* (St. Louis, Missouri: C. V. Mosby Co., 1981).

22. Moshe Feldenkrais, *Awareness through Movement* (London: Penguin Books, 1980), pp. 46–47.

23. See Juhan, *Job's Body*, Chapter Seven.

24. Lucille Daniels and Catherine Worthingham, *Muscle Testing* (Philadelphia: W. B. Saunders Co., 1980), p. 5.

25. Thompson, *Manual of Structural Kinesiology*, p. 3.

Chapter Eight

1. See, for example, Alexander Lowen, *Bioenergetics* (New York: Penguin Books, 1976), and Ken Dychtwald, *Bodymind* (New York: Pantheon Books, 1977).

2. Bonnie Bainbridge Cohen, Ruth Leeds, Linda Kaleb, Susan Peffley, and Kay Wylie, *The Skeletal System: Manual for a Workshop in Body-Mind Centering* (Amherst, Massachusetts: The School for Body-Mind Centering, 1977), p. 3.

3. See Joseph Campbell, *The Masks of God* (New York: Penguin Books, 1959–1968).

4. Bonnie Bainbridge Cohen, Patricia Bardi, and Gail Turner, *The Organs: Manual for a Workshop in Body-Mind Centering* (Amherst, Massachusetts: The School for Body-Mind Centering, 1977), p. 2.

5. Cohen, et al., *The Organs*, p. 2.

6. The work of Judith Kestenberg is interesting in this context. See, for example, *The Role of Movement Patterns in Development* (New York: Psychoanalytic Quarterly, Inc., 1967).

7. Cohen, et al., *The Organs,* p. 3.

8. Bonnie Bainbridge Cohen, "The Neuroendocrine System" (Amherst, Massachusetts: The School for Body-Mind Centering, unpublished), p. 9.

9. Cohen, et al., *The Organs,* p. 3.

10. *The Yellow Emperor's Classic of Internal Medicine* (Berkeley, California: University of California Press, 1972), Ilza Veith, translator; and Dianne M. Connelly, *Traditional Acupuncture: The Law of the Five Elements* (Columbia, Maryland: Center for Traditional Acupuncture, 1987).

11. Thorwald Dethlefsen and Rudiger Dahlke, *The Healing Power of Illness* (Dorset, England: Element Books, 1990).

12. See George Vithoulkas, *Homeopathy: Medicine of the New Man* (New York: Prentice Hall Press, 1987).

13. See, for example, Piero Ferrucci, *What We May Be* (Wellingborough, England: Turnstone Press, Ltd., 1982); and James Vargiu, "The Theory of Subpersonalities," *Psychosynthesis Workbook* (Palo Alto, California: Psychosynthesis Institute, 1974).

14. See, for example, Alice Bailey, *Esoteric Healing* (New York: Lucis Publishing Co., 1975).

15. The associations described here are adapted from Bonnie Bainbridge Cohen's "The Neuroendocrine System." Cohen's own work with the endocrine system has been particularly influenced by Alice Bailey's *Esoteric Healing*; it is, however, a different system. See also C. W. Leadbeater, *The Chakras* (Wheaton, Illinois, Madras, and London: Theosophical Publishing House, 1927); Douglas Baker, *Esoteric Anatomy* (London: Little Elephant, 1976); and Fritz Frederick Smith, *Inner Bridges* (Atlanta, Georgia: Humanics Ltd., 1986).

16. For an illustration of the coccygeal body, see Figure 377 in

Carmine D. Clemente, *Anatomy: A Regional Atlas of the Human Body, Third Edition* (Baltimore, Maryland and Munich: Urban & Schwarzenberg, 1987), plate 377.

17. Edwin B. Steen and Ashley Montagu, *Anatomy and Physiology, Volume 2* (New York: Barnes and Noble Books, 1959), p. 199.

18. The thoraco body and heart bodies were both discovered experientially by Bonnie Bainbridge Cohen. As yet, we have no concrete or scientific evidence available concerning the thoraco body. However, when we work with the perception of both of these structures in the body, we experience very powerful and distinctive energies located there.

19. See M. G. Nicholls, "Editorial and historical review," Mini-symposium: The Natriuretic Peptide Hormones, Introduction, in *Journal of Internal Medicine,* Vol. 235, 1994, pp. 507–514; and Harriet MacMillan and Meir Steiner, "Commentary: Atrial Natriuretic Factor: Does It have a Role in Psychiatry?," in *Biological Psychiatry,* Vol. 35, 1994, pp. 272–277.

20. See Walter Pierpaoli and Vladimir A. Lesnikov, "The Pineal Aging Clock: Evidence, Models, Mechanisms, Interventions," in *Annals New York Academy of Sciences,* Vol. 719, May 31, 1994, pp. 461–473.

21. Pierpaoli and Lesnikov, "The Pineal Aging Clock," p. 464.

22. See Pierpaoli and Lesnikov, "The Pineal Aging Clock," pp. 465–467.

23. See N. Vassiljev, J. Volyansky, V. Slepushkin, V. Kosich, and T. Koljada, "The Pineal Gland and Immunity," in *Annals New York Academy of Sciences,* Vol. 719, May 31, 1994, pp. 291–297.

24. Alan Bleakley, *Fruits of the Moon Tree* (London: Gateway Books, 1984), p. 187.

Chapter Nine

1. Stanley Keleman, *Living Your Dying* (Berkeley, California: Center Press, 1974).

2. Carl Sagan, *The Dragons of Eden* (New York: Ballantine Books, 1978), p. 43.

3. See Sagan, *The Dragons of Eden*.

4. Bonnie Bainbridge Cohen, "The Neuroendocrine System" (Amherst, Massachusetts: The School for Body-Mind Centering, unpublished), pp. 20–21.

5. Cohen, "The Neuroendocrine System," p. 36.

6. Robert E. Ornstein, *The Psychology of Consciousness* (New York: Penguin Books, 1975), Chapter Three.

7. Bonnie Bainbridge Cohen, "The Action in Perceiving," in *Sensing, Feeling, and Action: The Experiential Anatomy of Body-Mind Centering* (Northampton, Massachusetts: Contact Editions, 1993), p. 117.

8. Cohen, "The Action in Perceiving," p. 118.

9. Lennart Nilsson, *Behold Man* (Boston: Little, Brown and Co., 1974), p. 26.

10. Bonnie Bainbridge Cohen, quoted in Nancy Stark Smith, "Sensing, Feeling, and Action," in *Sensing, Feeling, and Action: The Experiential Anatomy of Body-Mind Centering* (Northampton, Massachusetts: Contact Editions, 1993), p. 65.

11. Cohen, from unpublished seminar notes, 1983.

12. Sid Gilman and Sarah Winans Newman, *Manter and Gatz's Essentials of Clinical Neuroanatomy and Neurophysiology, Seventh Edition* (Philadelphia: F. A. Davis Co., 1987), p. 39.

13. See Stanislav Grof, *Beyond the Brain* (Albany, New York: State University of New York Press, 1985), and *The Adventure of Self-Discovery* (Albany, New York: State University of New York Press, 1988).

14. See Madeleine Davis and David Wallbridge, *Boundary and Space: An Introduction to the Work of D. W. Winnicott* (London and New York: Penguin Books, 1983).

15. In fact the word "intelligence" has as its root in the Latin *intelligere,* one meaning of which is "to gather between."

16. Cohen, quoted in Lisa Nelson and Nancy Stark Smith, "The

Neuroendocrine System," in *Sensing, Feeling, and Action: The Experiential Anatomy of Body-Mind Centering* (Northampton, Massachusetts: Contact Editions, 1993), p. 62.

17. Theodor Schwenk, *Sensitive Chaos* (London: Rudolf Steiner Press, 1965), p. 13.

18. Deane Juhan, *Job's Body* (Barrytown, New York: Station Hill Press, 1987), p. xxii.

19. This is discussed in Nancy Stark Smith's interview with Bonnie Bainbridge Cohen, "Sensing, Feeling, and Action," in *Sensing, Feeling and Action,* pp. 63–65.

20. CSF, or a fluid similar to CSF, has been found within the fibers of connective tissue, according to R. F. Erlingheuser, "The Circulation of Cerebrospinal Fluid Through the Connective Tissue System" (Academy of Applied Osteopathy Yearbook, 1959), cited in Juhan, *Job's Body,* p. 73.

21. Fritz Frederick Smith, M.D. *Inner Bridges* (Atlanta, Georgia: Humanics New Age, 1986), p. 160.

22. Bonnie Bainbridge Cohen, "The Dancer's Warm-Up Through Body-Mind Centering," in *Sensing, Feeling, and Action: The Experiential Anatomy of Body-Mind Centering* (Northampton, Massachusetts: Contact Editions, 1993), p. 15.

23. John E. Upledger, *Craniosacral Therapy* (Chicago & Seattle: Eastland Press, 1983) and *Craniosacral Therapy II: Beyond the Dura* (Chicago and Seattle, Washington: Eastland Press, 1987).

24. Cohen, "The Dancer's Warm-Up," p. 15.

25. For a thorough description of fascia, see Ida Rolf, *Rolfing* (Rochester, Vermont: Healing Arts Press, 1989), Chapter Three.

26. The guidelines for the fluid exploration were created in this form by Body-Mind Centering teacher Gale Turner and myself for a class we co-taught in the training program of the School for Body-Mind Centering. This version of the guidelines is based on Bonnie Bainbridge Cohen's original material in "The Dynamics of Flow: The Fluid System

of the Body," in *Sensing, Feeling, and Action: The Experiential Anatomy of Body-Mind Centering* (Northampton, Massachusetts: Contact Editions, 1993). My thanks to Gale for permission to use it here.

Chapter Ten

1. Alan Bleakley, *Fruits of the Moon Tree* (London: Gateway Books, 1984), p. 50.
2. See Carl Sagan, *The Dragons of Eden* (New York: Ballantine Books, 1978).
3. In a lecture and reading given by Robert Bly for the London Convivium for Archetypal Studies, 1988.
4. Dianne M. Connelly, *All Sickness is Home Sickness* (Columbia, Maryland: Center for Traditional Acupuncture, 1986), p. 17.

Bibliography

References Cited

Assagioli, Roberto. *Psychosynthesis.* Wellingborough, England: Turnstone Press, Ltd., 1975.

Bailey, Alice. *Esoteric Healing.* New York: Lucis Publishing Co., 1975.

Baker, Douglas. *Esoteric Anatomy.* London: Little Elephant, 1976.

Barnes, Marylou R., Carolyn A. Crutchfield, and Carolyn B. Heriza. *The Neurophysiological Basis of Patient Treatment, Vol. 2. Reflexes in Motor Development.* Atlanta, Georgia: Stocksville Publishing Co., 1978.

Bartenieff, Irmgard, and Dori Lewis. *Body Movement: Coping with the Environment.* New York: Gordon & Breach Science Publishers, Inc., 1980.

Bleakley, Alan. *Fruits of the Moon Tree.* London: Gateway Books, 1984.

Bly, Robert. Lecture and reading given at the London Convivium for Archetypal Studies, 1988.

Campbell, Joseph. *The Masks of God.* New York: Penguin Books, 1959–1968.

Caplan, Frank. *The First Twelve Months of Life.* New York: Bantam Books, 1978.

Clemente, Carmine D. *Anatomy: A Regional Atlas of the Human Body, Third Edition.* Baltimore, Maryland and Munich: Urban & Schwarzenberg, 1987.

Cohen, Bonnie Bainbridge. *The Evolutionary Origins of Movement.* Amherst, Massachusetts: School for Body-Mind Centering, 1986.

—. "The Neuroendocrine System." Amherst, Massachusetts: The School for Body-Mind Centering, unpublished.

—. *Sensing, Feeling, and Action: The Experiential Anatomy of Body-Mind Centering.* Northampton, Massachusetts: Contact Editions, 1993.

Cohen, Bonnie Bainbridge, et al. "Joints and Ligaments." Unpublished manuscript. Amherst, Massachusetts: The School for Body-Mind Centering, 1982.

Cohen, Bonnie Bainbridge, Patricia Bardi, and Gail Turner. *The Organs: Manual for a Workshop in Body-Mind Centering.* Amherst, Massachusetts: The School for Body-Mind Centering, 1977.

Cohen, Bonnie Bainbridge, Ruth Leeds, Linda Kalab, Susan Peffley, and Kay Wylie. *The Skeletal System: Manual for a Workshop in Body-Mind Centering.* Amherst, Massachusetts: The School for Body-Mind Centering, 1977.

Connelly, Dianne M. *All Sickness is Home Sickness.* Columbia, Maryland: Center for Traditional Acupuncture, 1986.

—. *Traditional Acupuncture: The Law of the Five Elements.* Columbia, Maryland: Center for Traditional Acupuncture, 1987.

Daniels, Lucille, and Catherine Worthingham. *Muscle Testing.* Philadelphia: W. B. Saunders Co., 1980.

Davis, Madeleine, and David Wallbridge. *Boundary and Space: An Introduction to the Work of D. W. Winnicott.* London and New York: Penguin Books, 1983.

Dethlefsen, Thorwald, and Rudiger Dahlke. *The Healing Power of Illness.* Dorset: Element Books, 1990.

Dychtwald, Ken. *Bodymind.* New York: Pantheon Books, 1977.

Feldenkrais, Moshe. *Awareness through Movement.* London: Penguin Books, 1980.

Ferrucci, Piero. *What We May Be.* Wellingborough, England: Turnstone Press, Ltd., 1982.

Fiorentino, Mary R. *A Basis for Sensorimotor Development—Normal and Abnormal: The Influence of Primitive Postural Reflexes on the Development and Distribution of Tone.* Springfield, Illinois: Charles C. Thomas Publishers, 1981.

Gilman, Sid, and Sarah Winans Newman. *Manter and Gatz's Essentials of Clinical Neuroanatomy and Neurophysiology, Seventh Edition.* Philadelphia: F. A. Davis Co., 1987.

Grof, Stanislav. *The Adventure of Self-Discovery.* Albany, New York: State University of New York Press, 1988.

—. *Beyond the Brain.* Albany, New York: State University of New York Press, 1985.

Illingworth, Ronald S. *The Development of the Infant and Young Child.* New York: Churchill Livingstone, 1983.

The Illustrated Encyclopedia of the Animal Kingdom. Danbury, Connecticut: The Danbury Press, Grolier Enterprises Inc., 1968.

Juhan, Deane. *Job's Body.* Barrytown, New York: Station Hill Press, 1987.

Kapandji, I. A. *The Physiology of the Joints, Volume 3: The Trunk and Vertebral Column, Second Edition.* London and New York: Churchill Livingstone, 1974.

Kapit, Wynn, and Lawrence M. Elson. *The Anatomy Coloring Book.* New York: Harper and Row, 1977.

Keleman, Stanley. *Living Your Dying.* Berkeley, California: Center Press, 1974.

Kestenberg, Judith. *The Role of Movement Patterns in Development.* New York: Psychoanalytic Quarterly, Inc., 1967.

Leadbeater, C. W. *The Chakras.* Wheaton, Illinois, Madras, and London: Theosophical Publishing House, 1927.

Liang, T. T. *T'ai Chi Ch'uan for Health and Self-Defense.* New York: Vintage Books, 1977.

Lowen, Alexander. *Bioenergetics.* New York: Penguin Books, 1976.

MacMillan, Harriet, and Meir Steiner. "Commentary: Atrial Natriuretic Factor: Does It have a Role in Psychiatry?," in *Biological Psychiatry,* Vol. 35, 1994.

Man-ch'ing, Cheng. *T'ai Chi Ch'uan: A Simplified Method of Calisthenics for Health & Self-Defense.* Berkeley, California: North Atlantic Books, 1981.

Mills, Margret, and Bonnie Bainbridge Cohen. *Developmental Movement Therapy.* Amherst, Massachusetts: The School for Body-Mind Centering, 1979.

Moss, Richard. *The Black Butterfly.* Berkeley, California: Celestial Arts, 1986.

Nelson, Lisa, and Nancy Stark Smith, Interview with Bonnie Bainbridge Cohen, "The Neuroendocrine System," in *Sensing, Feeling, and Action: The Experiential Anatomy of Body-Mind Centering.* Northampton, Massachusetts: Contact Editions, 1993.

Nicholls, M. G. "Editorial and historical review," Minisymposium: The Natriuretic Peptide Hormones, Introduction, in *Journal of Internal Medicine,* Vol. 235, 1994.

Nilsson, Lennart. *Behold Man.* Boston: Little, Brown & Co., 1974.

Ornstein, Robert E. *The Psychology of Consciousness.* New York: Penguin Books, 1975.

Paxton, Steve. *Theatre Papers: First Series, Number 4.* Totnes, Devon, England: Dartington College of Arts, 1977.

Pearce, Joseph Chilton. *Magical Child.* New York: Bantam Books, 1980.

Pierpaoli, Walter, and Vladimir A. Lesnikov. "The Pineal Aging Clock: Evidence, Models, Mechanisms, Interventions," in *Annals New York Academy of Sciences,* Vol. 719, May 31, 1994.

Rolf, Ida. *Rolfing.* Rochester, Vermont: Healing Arts Press, 1989.

Sagan, Carl. *The Dragons of Eden.* New York: Ballantine Books, 1978.

Schwenk, Theodor. *Sensitive Chaos.* London: Rudolf Steiner Press, 1965.

Shaffer, Carolyn. "Dancing in the Dark," in *Yoga Journal,* November/December 1987.

Sheldrake, Rupert. *The Presence of the Past.* London: Collins, 1988.

Simonton, O. Carl, Stephanie Matthews-Simonton, and James L. Creighton. *Getting Well Again.* New York: Bantam Books, 1980.

Smith, Fritz Frederick. *Inner Bridges.* Atlanta, Georgia: Humanics Ltd., 1986.

Smith, Nancy Stark. "Sensing, Feeling, and Action: An Interview with Bonnie Bainbridge Cohen," in *Sensing, Feeling, and Action: The Experiential Anatomy of Body-Mind Centering.* Northampton, Massachusetts: Contact Editions, 1993.

Steen, Edwin B., and Ashley Montagu. *Anatomy and Physiology, Volumes 1 and II.* New York: Harper and Row, 1959.

Thompson, Clem W. *Manual of Structural Kinesiology.* St. Louis: Missouri: C. V. Mosby Co., 1981.

Upledger, John E. *Craniosacral Therapy.* Chicago and Seattle, Washington: Eastland Press, 1983.

—. *Craniosacral Therapy II: Beyond the Dura.* Chicago and Seattle, Washington: Eastland Press, 1987.

Vargiu, James. "The Theory of Subpersonalities," in *Psychosynthesis Workbook.* Palo Alto, California: Psychosynthesis Institute, 1974.

Vassiljev, N., J. Volyansky, V. Slepushkin, V. Kosich, and T. Koljada. "The Pineal Gland and Immunity," in *Annals New York Academy of Sciences,* Vol. 719, May 31, 1994.

Veith, Ilza, translator. *The Yellow Emperor's Classic of Internal Medicine.* Berkeley, California: University of California Press, 1972.

Vithoulkas, George. *Homeopathy: Medicine of the New Man.* New York: Prentice Hall Press, 1987.

Wilber, Ken. *The Atman Project.* Wheaton, Illinois: The Theosophical Publishing House, 1980.

Wolf, James M., editor. *Temple Fay, M.D.: Progenitor of the Domain-*

Delacato Treatment Procedures. Springfield, Illinois: Charles C. Thomas Publications, 1968.

Additional Reading

Asimov, Isaac. *The Human Brain.* New York: Mentor, 1965.

Ayres, Jane. *Sensory Integration and the Child.* Los Angeles: Western Psychological Services, 1979.

—. *Sensory Integration and Learning Disorders.* Los Angeles: Western Psychological Services, 1972.

Barlow, Wilfred. *The Alexander Principle.* London: Victor Gollancz, Ltd., 1990.

Bertherat, Therese, and Carol Bernstein. *The Body Has Its Reasons.* New York: Avon Books, 1979.

Blair, Lawrence. *Rhythms of Vision.* St. Albans, Hertsfordshire, England: Paladin Press, 1976.

Bobath, Bertha. *Techniques of Proprioceptive and Tactile Stimulation.* London: William Heinemann Medical Books Ltd., n.d.

—. *Techniques of Stimulating and Facilitating Spontaneous Movements.* London: Western Cerebral Palsy Centre, n.d.

Ferner, Helmut, and Jochen Staubsend. *Sobotta Atlas of Human Anatomy, Volumes 1 and 2, Tenth Edition.* Baltimore, Maryland and Munich: Urban & Schwarzenberg, 1983.

Freeman, William Harvey, and Brian Bracegirdle. *An Atlas of Invertebrate Structure.* London: Heinemann Educational Books Ltd., 1971.

Gray, Henry F. R. S. *Gray's Anatomy, Revised Fifteenth Edition.* Edited by T. Pickering Pick and Robert Howden. New York: Bounty Books, 1977.

Holle, Britta. *Motor Development in Children: Normal and Retarded.* Oxford, England: Blackwell Scientific Publications, 1976.

Ilg, Frances L., and Louise Bates Ames. *The Gesell Institute's "Child Behaviour."* New York: Dell Publishing Co. Inc., 1960.

Jolly, Alison. *The Evolution of Primate Behavior.* New York: Macmillan Co., 1972.

Kapandji, I. A. *The Physiology of the Joints, Volume 1, Fifth Edition; Volume 2, Second Edition, Volume 3, Second Edition.* Edinburgh, London, and New York: Churchill Livingstone, 1982, 1970, 1974.

Keleman, Stanley. *Emotional Anatomy.* Berkeley, California: Center Press, 1985.

Laban, Rudolf. *The Language of Movement.* Boston: Plays, Inc., 1974.

Laban, Rudolf, and F. C. Lawrence. *Effort.* London: MacDonald & Evans Ltd., 1974.

Laban, Rudolf, and Lisa Ullmann. *The Mastery of Movement.* Boston: Plays, Inc., 1971.

Nichols, David, John Cooke, and Derek Whiteley. *The Oxford Book of Invertebrates.* London: Oxford University Press, 1971.

Park, Glen. *The Art of Changing.* Bath, England: Ashgrove Press, 1989.

Piontelli, Alessandra. *From Fetus to Child: An Observational and Psychoanalytic Study.* London & New York: Tavistock/ Routledge, 1992.

Romanes, G. J. *Cunningham's Textbook of Anatomy, Tenth Edition.* London: Oxford University Press, 1964.

Romer, Alfred Sherwood. *The Vertebrate Story.* Chicago: The University of Chicago Press, 1959.

Samuels, Mike, and Nancy Samuels. *Seeing with the Mind's Eye.* New York: Random House Inc., 1975.

Sweigard, Lulu E. *Human Movement Potential.* New York: Harper & Row, 1974.

Todd, Mabel Elsworth. *The Thinking Body.* New York: Dance Horizons, 1972.

Torrey, Theodore W. *Morphogenesis of the Vertebrates.* New York: John Wiley & Sons Inc., 1962.

Warfel, John H. *The Extremities, Fourth Edition.* Philadelphia: Lea & Febiger, 1974.

—. *The Head, Neck and Trunk: Muscles and Motor Points, Fourth Edition.* Philadelphia: Lea & Febiger, 1973.

Watson, Lyal. *Supernature.* London: Coronet Books, 1974.

Whyte, Lancelot Law, Editor. *Aspects of Form.* London: Lund Humphries, 1968.

Wilentz, Joan Steen. *The Senses of Man.* New York: Thomas Y. Crowell Co., 1971.

Further Resources

A 1994–1995 directory of certified practitioners and teachers of Body-Mind Centering currently working in the United States, Canada, Australia, and various countries in Europe is available from the Body-Mind Centering Association, the professional organization for this work. Please contact:

Body-Mind Centering Association, Inc.
16 Center Street, Suite 530
Northampton, MA 01060
Phone: (413) 586-5971

For information about workshops led by Bonnie Bainbridge Cohen and professional Body-Mind Centering training programs led by Bonnie Bainbridge Cohen and other teachers, please contact:

The School for Body-Mind Centering
189 Pondview Drive
Amherst, MA 01002
Phone: (413) 256-8615
Fax: (413) 256-8239

Index